# ATLAS of the SOLAR SYSTEM

To Patrick
for setting me firmly on the road to space

■ FRONT ENDPAPER:
Voyager 1, its encounter with Saturn and its satellite system over, leaves the ecliptic plane away from the Sun, and takes a last look at the ringed planet as a crescent. By about 1990 Voyager 1 will leave the Solar System for ever, heading towards the constellation Ophiuchus. Meanwhile, on 25–6 August, 1981, despite jammed cameras, Voyager 2 returned even more detailed pictures of the Saturnian system before moving on to Uranus.

■ BACK ENDPAPER: The end of the Earth. It is five billion A.D., and the Sun almost fills the sky as a red giant. Humanity – in whatever form it has evolved – left aeons ago but one of its last almost indestructible cities still survives amid a cracked and barren desert.

# ATLAS of the SOLAR SYSTEM

Written and illustrated by
## David A. Hardy FRAS, FBIS

World's Work Ltd

Also by David A. Hardy

*Air and Weather*
*Energy and the Future*
*Light and Sight*
*Rockets and Satellites*
*The Solar System*

Designed by Victor Shreeve

Text and illustrations copyright
© 1981 by David A. Hardy
Index compiled by Ruth Hardy
Published by World's Work Ltd
The Windmill Press, Kingswood,
Tadworth, Surrey

Reproduced by Graphic Affairs Ltd,

Made and printed in Great Britain by
William Clowes (Beccles) Ltd,
Beccles and London

SBN 437 06540 5

# Contents

Introduction 7

Setting the Scene 8

The Dawn of Creation 12

The Sun Today 20

Earth Evolves 26

Man and the Universe 34

The Moon 40

Mercury 46

Venus 50

Mars 56

Cosmic Debris 64

Jupiter 68

Saturn 76

Uranus and Neptune 80

Pluto 84

Exploring the Future 86

Glossary 90

Table of Satellites 92

Index 93

# Introduction

By the end of this century almost all of the major planets may well have been investigated by unmanned space probes. The outer four, from Jupiter to Neptune, will all have been visited in turn by one probe, Voyager 2, while the Galileo probe to observe conditions in the atmosphere of the giant planet Jupiter is now scheduled for launch in 1985.

But this is a time of economic cut-backs. NASA has all but lost its opportunity to explore Halley's Comet, which makes its fly-by of our part of the Solar System in 1986, not to return for 76 years. NASA's Space Shuttle, though, by turning near-Earth manned space travel into an almost everyday affair, should lead to ventures further afield, using vehicles assembled in Earth-orbit. The use of the Space Telescope and the NASA/ESA Spacelab, both launched by the Shuttle, will also promote great advances in observational astronomy.

Interestingly, the present reduction in budgets for space exploration does not reflect a similar reduction in public interest. No doubt some people will continue to maintain that 'the money is better spent on Earth', (though far from being wasted by being blasted into space, it provides jobs for thousands); yet in January 1981 the newly formed US Viking Fund handed NASA its first contribution –$60,000– donated by ordinary people who felt strongly about the closing down of still-operable experiments on the Viking 1 lander. The Fund continues, with a growing number of contributors, and now there is a Halley Fund, led by eminent scientists and astronomers.

Meanwhile, the way is open for the amateur astronomer, and especially the planetary observer, to make valuable contributions. Some may have felt that their humble efforts were being eclipsed by the truly fantastic images and data sent back by the Mariner, Viking, Pioneer and Voyager probes. But these spacecraft return their results over a matter of days, months or, rarely, a few years. There is a great need for *continuous* observation: of the changes in position and rotation periods of the cloud belts and spots on Jupiter and Saturn; of the polar caps and dust-storms on Mars; of the Ashen Light of Venus; and of spots on the Sun itself. The amateur is ideally suited for this work.

So this book is intended not only as a timely retrospective review of Man's past discoveries and achievements in exploring the Solar System, but as a look ahead at the vast potential scope – practical as well as scientific – for his evolution in the future. It is necessary at this time for everyone concerned about space exploration to do all in their power to promote an interest in it; not least in the young, for youth needs to believe in goals for the future. There is a great, and understandable, absorption with today's problems on Earth, but this should not be allowed to cloud a worthwhile goal: Man's eventual expansion into the Solar System of which he is part. My hope is that this book will provide some inspiration to that end.

DAVID A. HARDY
Hall Green, Birmingham

# Setting the Scene

- Closest body to Earth: the Moon.
- Closest star to Earth: the Sun.
- Number of planets in Solar System: 9.
- Planets visible with naked eye: Mercury, Venus, Mars, Jupiter, Saturn.
- Number of stars visible with naked eye on a clear night: 2,000.
- Number of stars in Milky Way Galaxy: 100,000,000,000.

*'. . . Had we never seen the stars, and the sun, and the heaven, none of the words which we have spoken about the universe would ever have been uttered. But now the sight of day and night, and the months and the revolutions of the years . . . have given us a conception of time.'*

Timaeus PLATO (4th century BC)

Living amid the constant glare of modern city lights, many of us have lost that sense of wonder that first gripped our remote ancestors as they gazed up at the night sky. But if you can get away from house and street lights on some dark clear night and, after giving your eyes at least ten minutes to adapt, look up at the sky, you will be richly rewarded. No photograph, film, or written description can quite capture the strange beauty and remoteness of those scattered points of light. The nearest we can come to the reality, perhaps, is in a planetarium. What books, films and planetaria *can* do is to explain what is known about the stars and other heavenly bodies. In this book my aim is to tell the story of one of the stars – our Sun – and its family of planets, from their birth to their eventual death.

You could not know simply by looking up at the sky on one night that a few of those stars are really planets, or which; or that those planets are rocky or gaseous worlds circling our Sun and merely reflecting its light. One celestial body that is clearly visible and identifiable is our Moon. In the past its apparent ability to change its shape from night to night puzzled primitive astronomers. It was only by observing the sky over days, years and centuries that Man gradually built up a picture of the universe and the way it works.

With the invention and development first of optical instruments, then of radio and radar astronomy, and finally of unmanned and manned space probes and satellites, mankind has added enormously to his store of knowledge, so that we have learnt more in the last 30 years than in the previous 300 – or 30,000! Even so, there are many mysteries still to be solved, even in this tiny part of the universe we call the Solar System.

The Sun, the brightest object in the sky is, of course, the heart of our Solar System. The next brightest object is the Moon. By an amazing coincidence they appear almost exactly the same size in our sky and this led some ancient astronomers to assume that they were both the same distance from Earth. Yet apart from their apparent size, they have little else in common. Not only is the Sun far brighter, it also gives off a great deal of heat, whereas the far less luminous Moon is mottled with dark markings and it gives off no measurable heat of its own. The Moon also changes its position in the sky with relation to the Sun and stars in a regular monthly cycle. At the same time it goes through phases from New to Full and back – the Moon is, in fact, the only body whose phases can be seen with the naked eye.

'How vivid and radiant is the lustre of the fixed stars! How magnificent and rich that negligent profusion, with which they appear to be scattered throughout the whole azure vault! Yet if you take the telescope, it brings into your sight a new host of stars that escape the naked eye . . . is not the whole system immense, beautiful, glorious beyond expression and beyond thought?'
*Three Dialogues between Hylas and Philonous*
George Berkeley (1713)

It seems obvious to us today that these are due to the angle at which we look at the sunlit half of the spherical Moon, but to primitive Man they were inexplicable.

Once the telescope had been invented and put to astronomical use it became clear that some of the planets also showed phases. Before that, the planets were known only because, by repeated observations over months, they could be seen to change their positions (hence their name, from *planete*, the Greek word for 'wanderer'). Several of the planets appeared much brighter than the other stars, and so attracted attention. But all were observed to move along a fairly narrow path amongst the 'fixed' stars that the ancients had separated into arbitrary patterns or constellations, which they imagined resembled figures and animals. This path, with its constellations, is known as the Zodiac. (Were the stars visible in daylight, the Sun could be seen to move along the same track.)

To us, the very word 'planet' conjures up an image of another world, perhaps one on which Man might even land or live one day. To the ancients, though, the planets were regarded with curiosity and awe, many regarding them as gods. Their movements were carefully tabulated. It was soon discovered that the planets (only five of which were known: Mercury, Venus, Mars, Jupiter and Saturn) did not move steadily – as did the Sun and Moon – but sometimes speeded up or slowed down, or even reversed direction completely. To early thinkers such as the Sumerians living 6,000 years ago in what is now Iraq, this must have been a challenge. They liked an orderly universe, and were probably the first to group the stars into constellations. Their further attempts to derive some sort of pattern from the erratic wandering stars may have formed the basis of astronomy as we know it.

Much of what the ancients saw in the sky led them to conclude that the Earth was the centre of the Solar System, and, indeed, the whole Universe as they understood it. The Sun rose in the east, crossed the sky and set in the west; 'obviously', then, it went round the Earth. So did the Moon (and this time the statement was correct). The stars and planets, too, as they discovered, rose and set. Further careful observation of all these bodies showed that they rose and set at a slightly different point on the horizon each night (or day), but that they did eventually reappear at the original point over a long period, and that the cycle is repeated over and over. We now believe that circles of 'standing stones' such as Stonehenge assisted such observations.

The rising and setting of the Sun gives us our 'day'. The Moon's longer period from New to Full and back provides the 29.5-day 'month', while the Sun's return to a given point led to the 'year', at first estimated at about 360 days. Through the ages there have been many 'calendars' invented to measure these time-scales; the earliest examples – notches on sticks, as used by some aborigines today – date back to Palaeolithic times, around 30,000 B.C.

It took a very long time for the model of the universe which we now take for granted to be generally accepted. It is not the Sun's movement around Earth that causes it to rise and set, but Earth's own rotation of 24 hours. Earth takes 365.25 days to return to the same point in its almost circular orbit around the Sun. Similarly, more and more of the Moon's face is illuminated as it revolves around Earth, away from the Sun, then spins slowly into darkness again.

No great feat of deduction was needed for the ancients to assume that the Sun is 'burning' – though the nature of that fire took a lot longer to determine. But what of those tiny, remote sparks of light, the stars? How could they have guessed that they were in fact suns too – many of them vastly larger and brighter than our own? (As a matter of fact just such a suggestion was made around 1450 by a German archbishop, Nicholas of Cusa. He added that each might have its own retinue of worlds on which intelligent life might exist; this was over 150 years before the telescope was invented – he was ignored, of course.)

Speculation about the Sun, the Solar System and the stars beyond continued for centuries. Modern astronomy and the use of sophisticated technology has enabled us to build up a coherent picture of our Solar System and its place in the wider universe:

The star we call the Sun is just one of approximately 100,000 million suns forming a spiral-shaped galaxy which we call the Milky Way (from the pale band of light it forms in our night sky). The Sun is an insignificant member of the galaxy, occupying a position in one of the spiral arms; it is larger and brighter than some, but smaller and dimmer than many other stars.

The Milky Way is in turn one of what is called the local group of over two dozen galaxies (not all spiral), but beyond them are millions of other galaxies – each containing a similar number of stars. There is no reason to suppose that

■ TOP: As Earth overtakes slower-moving Mars in its orbit, the red planet seems to reverse its direction as seen against the background of 'fixed' stars. It would take several months to plot the path shown here. ■ RIGHT: The heavenly bodies move across the apparent inverted bowl of the sky. This means that the Moon often appears to reflect light from a different angle to that at which the Sun's rays must be striking it. At Crescent phase, a straight line drawn at a right angle to another line connecting the horns will pass through the Sun. Beyond First Quarter one has to use a curved line to allow for the 'bowl' illusion. (The diagram is drawn as seen from the northern hemisphere.)

■ LEFT: The orbits of the planets to scale. **1**. Mercury **2**. Venus **3**. Earth **4**. Mars **5**. Asteroid belts **6**. Jupiter **7**. Saturn **8**. Uranus **9**. Neptune **10**. Pluto. Note that the orbit of Pluto crosses that of Neptune; there is no danger of collision as Pluto's orbit is highly inclined.

■ ABOVE LEFT: The approximate position of the Solar System in a spiral arm of the Milky Way Galaxy. ABOVE RIGHT: An 18th-century orrery. One of the earliest of these instruments was made for the Earl of Cork and Orrery – hence the name used today. The planets are made to revolve around the Sun at their correct relative periods, but greatly speeded up, using an ingenious system of gears. They are the forerunners of today's planetaria.

■ RIGHT: The planets to scale with the Sun.

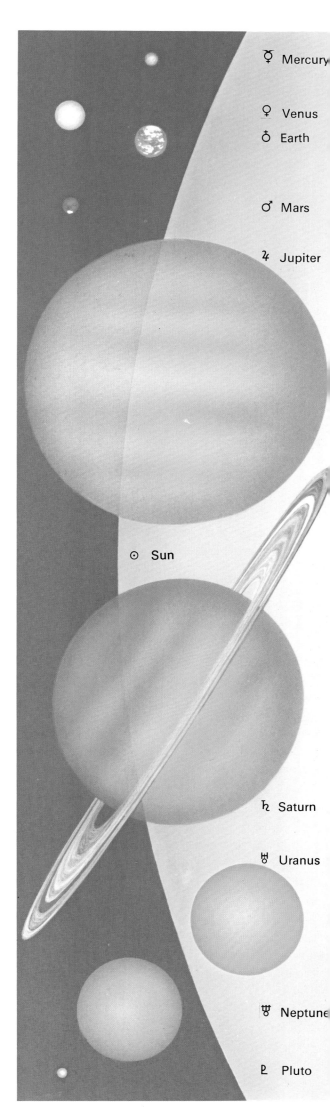

☿ Mercury

♀ Venus

♂ Earth

♂ Mars

♃ Jupiter

☉ Sun

♄ Saturn

♅ Uranus

♆ Neptune

♇ Pluto

a great many Sun-like stars in each galaxy do not have habitable planets in orbit around them, though none of our instruments have yet been able to prove this, even in our own galaxy.

Round our Sun circle (at least) nine planets. Their orbits are actually not circles but ellipses, and all but Pluto's are roughly in the same plane, though their axes of rotation vary considerably.

The inner planets, from Mercury to Mars, are basically rocky, like the Earth; only Mercury has no appreciable atmosphere. Then, in a wide gap between the orbits of Mars and Jupiter are about 2,000 known (and possibly 100,000 as yet uncharted) bodies called asteroids (meaning 'starlike'). Their diameters vary from 1,000km to a few hundred metres or less, so that they are sometimes given the more accurately descriptive names planetoids or minor planets. Despised at one time by astronomers as 'vermin of the skies', these bodies could well hold the key to Man's industrial future.

Beyond the asteroids, separated from each other by hundreds of millions of kilometres, are the orbits of the outer planets. The outermost is Pluto, which is probably a small (smaller than Earth or Mars) frozen ball of ice. Apart from this 'odd man out' the outer planets are giants. Closest to us, beyond the asteroids, is Jupiter, the largest of all – over 10 times the diameter and 1,300 times the volume of Earth. Next comes Saturn, until recently believed to be the only planet with a system of rings around it. We now know that both Jupiter and the next planet out, Uranus, are ringed, and it seems probable that Neptune will prove to have rings too.

Most people will at some time have seen a streak of light flash across the night sky and silently fade, almost as though a star had left its place and dived to Earth. This is exactly what the ancients thought had happened. Their origin lies in the fact that space is full of orbiting debris ranging in size from grains of dust to 'flying mountains' and known as meteoroids. When a meteoroid drifts into the clutching fingers of Earth's gravity it is pulled into the atmosphere, where its very high speed causes it to become incandescent as a meteor and – unless it is very large – vaporise. The larger ones reach the ground as meteorites and some strike with such an impact that they may even cause a crater and considerable damage. Reports of stones falling to the ground from an empty sky were for many centuries flatly disbelieved, and in any case not connected with 'shooting stars'.

Other occasional visitors to Earth's skies are comets. These celestial 'ghosts' trail their tenuous veils of dust and gas millions of kilometres from Earth as they approach from the outer fringes of the Solar System, swing around the Sun and retreat into outer space, some never to return.

Most comets describe highly elliptical orbits, that bring them back to our neighbourhood in anything from a few to thousands of years. Although each comet has its own character, in all cases the tail points away from the Sun (contrary to popular misconception, a comet does not shoot across the sky with its tail streaming out behind). As it approaches the Sun the tail is behind, but as it swings round the Sun the comet's tail is in front of its head! Those in the vicinity of Earth remain visible for many nights, slowly changing position against the stars.

# The Dawn of Creation

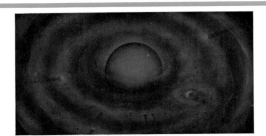

- Most abundant element in Universe: Hydrogen.
- Age of the Universe: 10–20,000,000,000 years.
- Age of the Sun: 5,000,000,000 years.
- Age of the Earth and planets: 4,600,000,000 years.
- Age of oldest rocks found on Earth: 3,800,000,000 years.

*'Consider the excellence of the sun, prime prince and controller of the world, favouring and forwarding every life that is. By its chastened heat it supports all living things in their doing what they do.'*

De Abditis Rerum Causis JEAN FERNEL (1548)

Very occasionally astronomers are able to record the appearance of a brilliant star in the sky where none had been seen before. It remains for a few days, reaching a peak of brightness, then slowly subsides, perhaps over years. More rarely and even more spectacularly an even brighter star is recorded. These phenomena are known as novas and supernovas from the Latin *stella nova*, meaning 'new star', which is what the ancients took them to be. Only comparatively recently has it been found that they represent not the birth but the violent death throes of a star. Curiously, the latest theories suggest that it was a supernova that was responsible for the birth of our own Sun, and hence the Solar System and Earth.

The first serious attempt at a theory on the origin of the Solar System came in 1796 when the French astronomer Pierre de Laplace advanced his 'Nebular Hypothesis'. It actually originated with the German philosopher Immanuel Kant, but Laplace had improved on it. The idea was that a great disc-shaped cloud of gas, slowly rotating, cooled and shrank, radiating its heat into space. As it did so its rate of spin increased until the centrifugal force at its edge equalled the gravitational pull; a ring of matter broke away and slowly condensed into a planet.

This continued several times until a central Sun remained, with circling planets in attendance. It was a theory that was generally accepted for many years. However, better mathematics showed that the present rotational motion (known by astronomers as angular momentum) of the Solar System could not possibly have been produced by this method, and anyway such a contracting cloud could never have rotated fast enough to throw off such rings.

On the other hand, the outer rings of the cloud (which was vastly larger than the present diameter of the Solar System) could never have condensed into planets if they originally had as much rotational motion as the planets do *today*.

The total angular momentum of a system of bodies is calculated by adding together their individual angular momentums. That of each planet remains constant throughout its orbit; this is an important factor. The Sun rotates once every 25 days, so its own angular momentum is a very small fraction of all the other planets.

Had Laplace been correct, the angular momentum of the system would be concentrated mainly in the Sun itself; in fact, nearly all of it is concentrated in the four 'giants' – Jupiter, Saturn, Uranus and Neptune. This was a great puzzle, since no changes *within* a system can alter its angular momentum – yet most of this should have remained in the contracting core of the original primeval cloud. Clearly a new theory was needed!

His (Aristarchus') hypotheses are that the fixed stars and the sun remain unmoved, that the earth revolves about the sun in the circumference of a circle, the sun lying in the middle of the orbit.
Archimedes, writing to King Gelon of Syracuse (5th century BC)

In an attempt to find a solution to this problem two Americans, an astronomer, F. R. Moulton, and a geologist, T. C. Chamberlin suggested in 1900 that in the remote past another star had passed close to the Sun, drawing from it great streams of gas and matter. The cloud of material then went into orbit around the Sun, cooled, and solidified, forming 'planetesimals' that gradually accumulated into planets. Other scientists formulated different versions of this theory. One, by Sir James Jeans, was that the gravity of the passing star pulled a long, cigar-shaped filament from the Sun, with Jupiter and Saturn condensing at the thickest part and smaller planets at each end – as is, of course, the case.

Another hypothesis was that the Sun was once a member of a double-star system, or binary star, many of which exist. Either the companion star was struck by an intruder, or it was merely torn away from its mutual orbit with the Sun, in either case leaving behind a trail of planet-forming debris whirling around the Sun.

It was Professor (now Sir) Fred Hoyle who suggested that the Sun's companion had exploded with great violence as a supernova. The explosion was sufficient to hurl it out of the system altogether, ejecting a cloud of gas as it went which condensed in a way that gave rise to planetesimals.

An alternative to these 'catastrophic' theories was a further development of the less violent 'evolutionary' theories of Kant and Laplace. It came from a German physicist, Carl von Weizsäcker, in 1943, and was developed in the 1950s by American astronomer Gerard P. Kuiper, and others.

They argued that while space may appear to be empty – an almost perfect vacuum compared to anything we can produce in a laboratory – it is actually filled with gas, mainly hydrogen and helium. The atoms are very widely spaced; yet denser regions must exist and the early Sun or 'protosun' was formed in, or passed through, one of these, collecting a huge gaseous envelope. This became a rotating, contracting disc of gas and particles, which gravitational effects and random swirling motions again formed into clumps or planetesimals.

■ RIGHT: The primordial nebula, illuminated by a group of young, rapidly-burning stars. One of these is flaring into a supernova, and its pressure wave and cooling mineral grains have a profound effect upon the cloud (inset). Such a group of stars could trigger a whole wave of star formation, the new stars then producing *more* supernovae and so on until all the material in that vicinity is used up. (See also overleaf.)

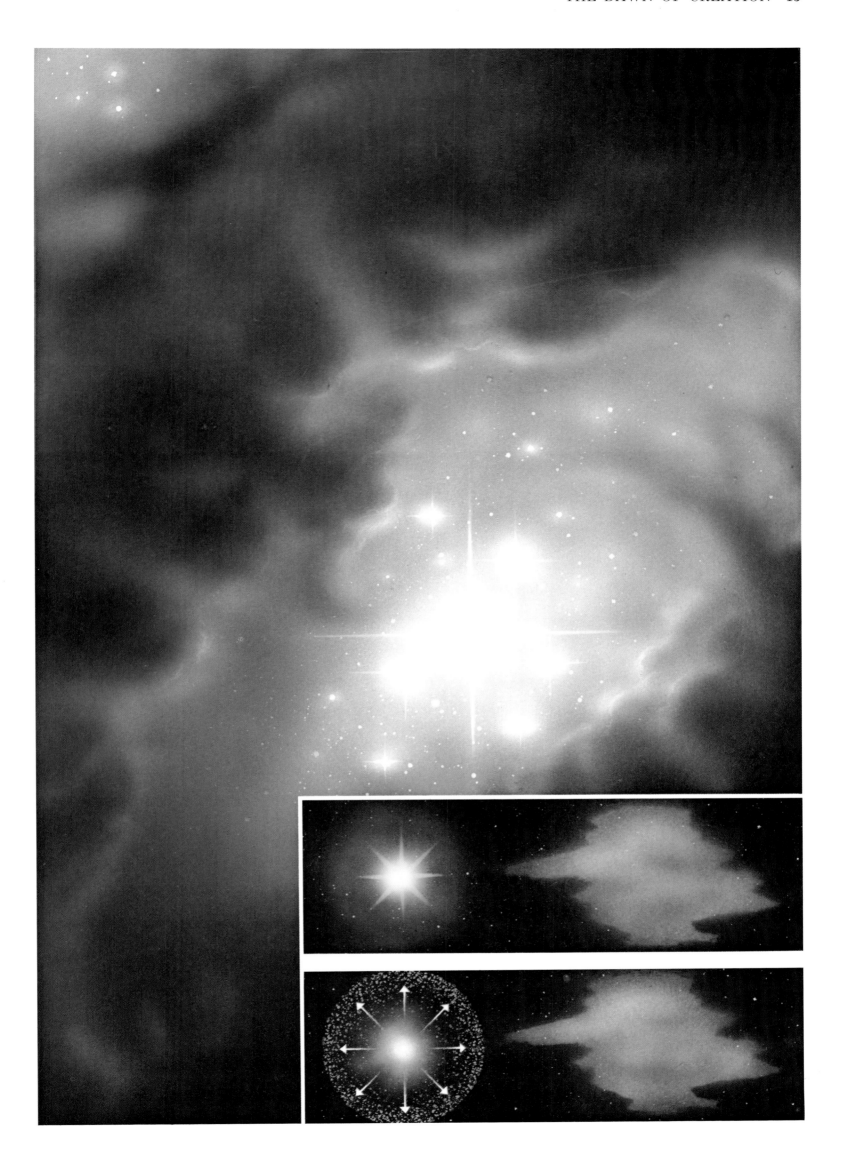

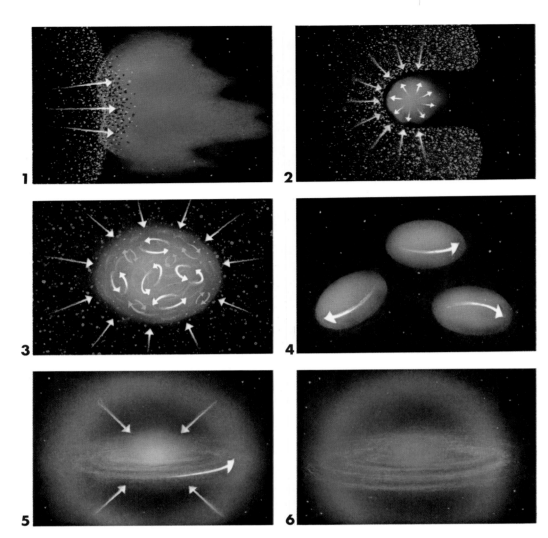

■ LEFT: The protosolar cloud is penetrated by grains from the supernova **1**, while it is compressed by the gas from the explosion, which eventually surrounds the cloud **2**. When the cloud achieves a high enough density it collapses under its own gravity **3**, but this creates violent eddies that cause it to break into fragments **4**, each with its own rotation. The central portion of one of these fragments collapses more rapidly into a central mass surrounded by a flattened, spinning disc of gas and dust grains with an outer shell composed of more gas and dust **5**. Eventually the protosun forms at the centre of a disc-shaped, rotating cloud, continually growing by accretion **6**.

Apart from the gases, there were a few dust particles consisting of heavier elements. These formed less than 5 per cent of the total mass of the cloud, and were so scattered that only 100 such grains would be found in a cubic kilometre. But they were important, because they contained all the elements from which our world is made, such as silicon, iron, aluminium, etc. There were also organic molecules composed of carbon, oxygen and nitrogen – long before there were even planets for life to exist on, or a Sun to warm them.

As long ago as 1644, René Descartes had suggested that rotational eddies occurred naturally in a primordial gas which filled the universe, and even Kuiper believed in similar random movements. Today most cosmogonists believe that some external force was needed to disturb the cloud. Researchers are far from agreed on what this was, but two or three main hypotheses emerge – and in spite of the difficulty in obtaining observational support, some surprising new evidence is aiding at least one of these.

We know that our galaxy, the Milky Way, is well over twice as old as the Solar System, so it would not have been very different when our Sun and planets formed. Today we see great gas-clouds, some glowing with hot young stars, some dark with interstellar dust, in the spiral arms. In these, new stars may still be forming. It could be that one of these arms (see illustration on page 11) passed through our region of space 4.6 billion years ago, compressing the gas-cloud and disturbing its equilibrium.

The density of the dust grains would be increased perhaps a hundredfold, blocking the light and radiation from neighbouring stars, just as we see happening in the dark clouds in our galaxy and others. This caused the temperature in the primordial cloud to plummet, which in turn lowered the pressure of the gas. Internal pressure was thus no longer able to counteract gravity, and the cloud started to contract.

Because it would not contract uniformly, eddies could now form and the cloud would begin to break up, forming as many as several hundred protostars, each surrounded by its rotating disc of

As more and more mass was concentrated in the protosun its rate of rotation increased (like an ice skater who draws in his or her arms while spinning); but friction caused angular momentum to be transferred constantly from the rotating core to the disc. The disc was thus made to rotate at a very high speed (several thousand kilometres per second).

As the core contracted it became hotter until it radiated energy, which was absorbed by the disc. Naturally the inner part of the disc became hotter than the outer, meaning that its atoms and molecules moved at much higher speeds. High enough, in fact, for the inner disc to lose its lighter atoms almost completely. The pressure of radiation and particles emitted by the now glowing protosun swept the lighter elements towards the edge of the disc. The heavier elements that remained eventually formed the rocky inner planets, while the lighter elements were concentrated in the four 'gas giants' at the edge of the system.

Although an oversimplification, the basic mechanism of the theory should be clear. This does not, however, mean that all astronomers now agree on how our planetary system was formed, or even that the whole story is told above.

Apart from the normal human curiosity, and a thirst for knowledge, there is another important reason for discov-

ering precisely *how* the planets were formed. The fact is that supernovae or even collisions between stars are, as far as has so far been observed, rare. Thus, if our system of planets was created as a result of one of these, then we cannot expect to find many other such systems, even amongst the billions of stars in our galaxy. Interstellar gas-clouds or nebulae, on the other hand, are quite common, and evolutionary theories seem to suggest that planets might be a natural part of the scheme of things, appearing wherever conditions are suitable. There are still almost as many theories today as there are cosmogonists (cosmogony is the study of the origins of the Solar System). The best-accepted modern theories differ at various points, but most agree on the early stages:

## Creation Theories

Five billion years ago the space now occupied by our Solar System was filled with a vast cloud of gas and dust; almost 75 per cent by weight of the gas was hydrogen and almost 25 per cent helium, plus some neon. The atoms of the gas mixture were widely separated – a matchbox full would contain only a few dozen atoms, compared with several hundred million *trillion* atoms when full of our air at sea level. It would have been cold; less than 50 degrees Kelvin (−223°C. The Kelvin scale starts at Absolute Zero – minus 273°C).

gas and dust. One of these became our Sun.

Alternative theories suggest that the instability was caused by the formation of massive stars near by, or by rearrangements of the interstellar magnetic field. But the hypothesis which seems to be gaining more favour today takes us back to the supernova theory (though not as a member of our system), and oddly enough it is backed by evidence from the most humble members of the Solar System – meteorites.

In 1969 a two-tonne meteorite fell near the village of Pueblito de Allende in Mexico. Close examination of its fragments showed it to be a carbonaceous chondrite, a class of meteorite believed to be composed of the earliest materials to solidify from the primordial solar nebula. But when it was examined by three scientists at the University of Chicago, R. Clayton, T. Mayeda and L. Grossman, it was found to contain a very pure isotope of oxygen, O-16 (an isotope of an element is a form that is identical in all other ways except that it contains a different number of neutrons in its nucleus).

## Allende Meteorite

Although O-16 is found in terrestrial rocks, it is always in combination with isotopes O-17 and O-18. When the first rocks came back from the Moon in the same year, they too were found to contain basically the same ratio as Earth rocks.

A 'normal' star produces several isotopes of each element – iron, silicon, carbon, etc. In the case of oxygen there would be a mixture of O-16, O-17 and O-18.

The only known way for the almost pure O-16 to have found its way into the Allende meteorite (and others examined since) is if the oxygen came from a supernova explosion, in which most of the star's material is blasted into space. Any O-17 and O-18 would be destroyed in the intensity of the nuclear reactions that caused the outburst. To clinch the matter, it seems, it was then found that the unusual proportion of isotopes of magnesium-26 and alumin-

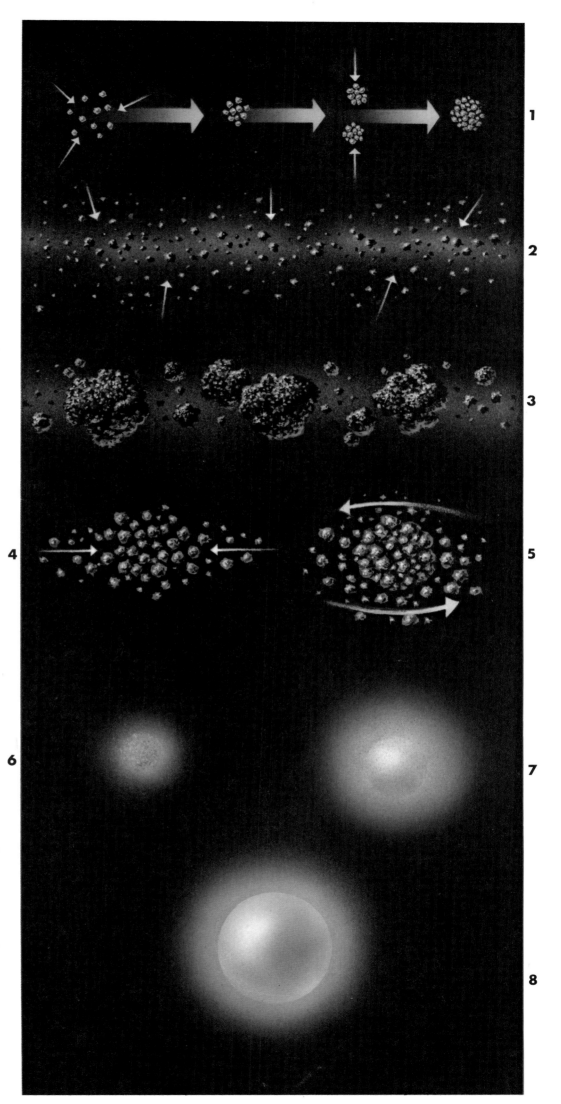

■ 1. The fluffy dust grains collide and stick, forming ever-larger clumps.
2. The clumps tend to congregate in the mid-plane of the nebula, forming a diffuse disc.
3. Mutual gravity causes them to collect into asteroid-sized bodies.
4. As clusters collide, their gravity fields relax.
5. They form dense bodies, with smaller bodies in orbit.
6. Finally a planet-sized body is formed.
7. The gravity of the new body causes gas from the nebula to be concentrated around it.
8. If the core is large enough, the gas will collapse into a dense shell.

ium-26 indicate that they, too, were affected by a supernova explosion.

It is estimated that the maximum distance of the supernova from the cloud to be effective would be 60 light years (a light year is the distance light travels in a year at nearly 300,000km/s). The material ejected would be immensely hot gas, but grains of minerals would probably condense out quite quickly as the shock wave expanded. When this reached our gas-cloud some of these grains would penetrate; these would contain the unique isotopes of oxygen, magnesium and aluminium, which we now recover in meteorites.

Having accepted the primordial cloud theory, do we now have to revert to a catastrophic origin of solar systems? Certainly, observations of the remnants of known supernovae reveal little patches of gas and dust that could well be stars forming.

But this does not necessarily mean that *all* stars or planetary systems have to be created in this way; the other processes mentioned may operate also, or still others as yet unknown.

There is a class of star known as infrared stars, which are either invisible or only faintly visible to our eyes but show up clearly on photographic emulsion sensitive to radiation at the infra-red ('below-red') end of the spectrum. These appear to consist of a young star inside a cloud of gas containing dust grains at quite low temperature – a few hundred degrees Kelvin – which block most of the star's visible light. When the light from these stars is broken down and examined it shows that some of these

grains are composed of silicates; that is, they are rocky. There seems little doubt that when we observe these 'cocoon nebulae' we are witnessing the early stages in the birth of new solar systems.

This stage, with gas and dust grains swarming around a central core, is not so different from those earlier theories. But whereas they assumed that all the material which formed the planets was hot and molten, like lava from a volcano, the particles in the cocoon nebula are quite cold. (A carbonaceous chondrite, as evidence of this, contains water bound into its substance which would have boiled away if the rock had ever been very hot.)

As the grains coalesce into larger bodies and others evaporate or are blown out of the system by radiation pressure from the central star, the light from the star can at last shine through and it becomes visible from outside.

We now have enough pieces of the puzzle to try to reconstruct what might have taken place in the cloud that became our Solar System.

Initially, its atoms fell naturally toward the centre of gravity, the core. As the atoms became more densely packed they began to collide and so created pressure, which acted against the pull of gravity. Such contraction in which pressure outward opposes an inward collapse causes an increase in internal temperature. (It is sometimes known as 'Helmholtz Contraction' after the German astrophysicist Hermann von Helmholtz). In the core atoms would also collide as they were pulled toward its centre, creating heat. As the heat in-

creased the protosun began to radiate, at first in the infra-red.

More and more of the radiation from the protosun, which previously could escape into space, was now absorbed by the increasing concentration of matter in the nebula, and the temperature rose. Gases which were frozen onto dust grains began to evaporate.

Within only a few thousand years the radiation became visible as light, as the temperature reached several thousand degrees K, but it was not yet a 'true' star. If the mass of the cloud were a lot smaller than that of the Sun it would

■ ABOVE: An imaginary scene inside the protosolar nebula. Planetesimals up to several kilometres across swarm around the protosun. Some drift together, but others collide at high speed (right), creating a spray of smaller particles and either shattering the large bodies or welding them together. ■ BELOW: 'Helmholtz Contraction' is the name given to the process whereby heat is created when outward pressure opposes an inward collapse. The German astrophysicist Hermann von Helmholtz first showed in 1871 how gravitational contraction would cause a rise in internal temperature (in 1847 he had formulated his famous Law of Conservation of Energy.) As the protosun shrinks it gives off more and more radiation.

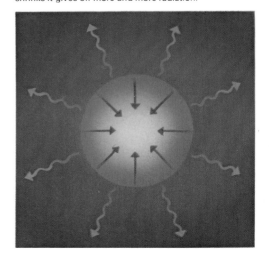

merely cool, becoming a 'black dwarf'. However, in the case of our solar cloud, the temperature continued to rise until it reached 10,000–20,000K at the centre; gas pressure again balanced gravity and the collapse was almost arrested.

The protosun, by now disc-shaped, was so massive at this stage, that it could radiate only from its surface – heat being brought there from inside by convection currents. The contraction continued, but more slowly and irregularly. The more it contracted the faster its rotation became, until it assumed a 'flying saucer' shape, with a central bulge.

In the first thousand years the gravity acting on the core drew material into the centre, making the core contract so that it shrank from a vast cloud to some 20 times the size of today's Sun. It also became 100 times brighter than our Sun for a brief period – known as its high-luminosity phase – and continued its contractions for anything up to 10 million years, slowly fading in brilliance all the time.

We have already seen how the outer parts of the disc speeded up at the expense of the centre. Material spiralling into inner orbits eventually caused the protosun to grow by accretion.

Meanwhile, at the outer edges of the disc where temperatures were lower the dust grains were undergoing change. Even before the protosun began to glow, probably, mutual gravity was causing them to collide and stick together, also growing by accretion. Atoms of gas may have reacted, forming molecules on the surfaces of grains. Acting as nuclei, the grains became centres for condensing substances.

This process continued after the protosun faded from its high-luminosity phase, with different precipitates condensing in various parts of the nebula and at differing temperatures. First metallic oxides, then metals such as nickel-iron, then silicates and various minerals would all condense out at decreasing temperatures. The ever-present hydrogen would combine with other gases to become water, ammonia and methane, in the form of frozen crystals in the coldest outer regions.

## Minute Beginnings

In addition, the tiny grains may have become coated with sticky, organic hydro-carbons forming fluffy, insubstantial, snowflake-like globs that easily adhered to each other and aided the formation of clumps several millimetres or even centimetres across, growing all the time. These tended to congregate in the mid-plane of the disc. As the years passed in their millions, the fitful bursts of radiation from the protosun caused nearer aggregations to fuse together, then solidify, until some became big enough to attract more matter to themselves by gravity alone.

Small planetesimals grew to the size of asteroids, which continued to collide, split, merge, melt together and finally formed a few large protoplanets. The four inner planets accreted mainly rocky materials, together with some metals that had been volatilized by the high temperature near the protosun.

The four large outer planets (ignoring Pluto for the moment), of course, received many of the gases, chiefly hydrogen, helium and neon, blown from the inner regions. Because of the huge radius of their orbits they would naturally sweep up more material than the inner planets – and the more they grew the more they collected. Being colder, the dust grains that formed their cores were layered with 'ices': water ice, carbon dioxide and, farther out, ammonia and methane.

An additional internal heating effect in the growing protoplanets was contributed by the radioactive decay of isotopes such as aluminium-26. These were formed by violent nuclear reactions inside the protosun, and were ejected as it went through its birth pangs in the process of becoming a true star. The interiors of protoplanets which intercepted these melted due to the heat released by such short-lived elements as aluminium-26 and also of uranium ores, which have a much longer 'half-life' (that is, the time taken for half of any quantity of an element to change, by radioactive decay, into a different element. It enables accurate dating of rocks anywhere in the universe).

Heavier elements such as iron then sank to the centre of the protoplanets, which in turn forced lighter materials upwards, nearer the surface. This process, known as chemical differentiation, was an important one in the formation of the planets as we know them today. It resulted in dense metallic cores – mainly iron and nickel – surrounded by rocky layers.

One important phase in the evolution and growth of the inner planets, which could be theorised and suspected by observing our Moon but not proved until the advent of space probes such as Mariner, Pioneer and Voyager, is their continued accumulation of smaller bodies by gravitational attraction. The evidence of the last phase of this can be seen in the cratering of Mercury, Mars and the satellites of Jupiter and Saturn. This constant bombardment also contributed external heating effects.

Whereas an asteroid-sized body would cool within a few million years, larger planets which melted internally and externally as they grew required much longer, and indeed most still are molten at their cores. When the terrestrial planets were still in the molten state, gases were released and either formed temporary atmospheres or escaped into space.

## Planets Take Shape

Mercury, being small (low gravity) as well as close to the solar furnace, quickly lost its gases. Mars is also small, but being farther from the Sun has retained a thin atmosphere of carbon dioxide. Of the inner planets only Venus and the Earth have been able to keep a protective layer of gases – though Earth's, especially, was to undergo great changes over the ages.

Beyond Mars a swarm of planetesimals seems to have failed to coalesce into one planet, and the asteroid belts were formed instead.

Although hydrogen and helium are the lightest gases, Jupiter and Saturn grew so large that they were soon surrounded by dense clouds of these gases, along with ammonia and methane and a few others. The low temperature helped to prevent these light gases from escaping.

But the giant planets also acted as miniature 'protosolar systems' themselves. Each of their gravitational whirlpools attracted its own disc of gas, dust and planetesimals which, instead of smashing into Jupiter and Saturn to add to their growth, went into orbit and formed a family of satellites. Uranus and Neptune were similar but failed to grow as large as the other two gas giants, and attracted fewer moons. One of Neptune's may later have deserted to become Pluto.

Far beyond Neptune's orbit, small fragments of the original primordial nebula may now have been in orbit around the main protosolar cloud. By processes similar to those that created the planets, they formed a vast shell of icy comets.

On page 14 we saw how friction between the protosun and the cloud caused the Sun to lose its initial rapid rotation, which had given rise to its disc shape. The Swedish-American astronomer H. Alfvén pointed out around 1960 that the protosun would probably have had a strong magnetic field, and would also shoot out ionised particles of gas – that is, electrically charged atoms. (Normally atoms are electrically neutral: if electrons are stripped from an atom it becomes a positively charged ion; if added, negatively charged.)

The ions in the inner part of the cloud would be trapped by the magnetic field, and would be dragged around with it as the protosun rotated, acting as a brake. As the early Sun's spin was slowed and its angular momentum transferred to the nebula, its shape became ellipsoid and finally spherical as we see it today.

All this time the pressure within the protosun had continued to grow as trillions of tonnes of gas were sucked into its gravitational maw, and its core grew steadily hotter. At last, some 4.5 billion years ago, the temperature there reached about 10 million degrees K. At this immense temperature atoms of hydrogen fused into helium, and fusion reactions started. Our Sun had ignited.

Stellar observers study an interesting class of stars which are similar in mass to the Sun, but whose internal fusion reactions have apparently not yet steadied down to the stable state that will put the stars into what is known as the 'main sequence' of stellar evolution, as is the Sun. These rapidly rotating stars fluctuate in brilliance (that is, they are variable stars); their light can vary by 50 per cent over a matter of hours or days, but on average they are from five to ten times as bright as they will be when they reach equilibrium. They are known as T-Tauri stars, from the variable star of this type first discovered in Taurus in 1943.

Usually these are found in or near one of the great glowing gas-clouds such as the Orion Nebula in our galaxy which are in all probability the birthplace of new stars. Often a T-Tauri star is also surrounded by a cocoon nebula, suggesting that a planetary system is in the process of forming around it, but the older stars appear to be losing mass rapidly as it becomes a 'stellar wind' flowing away from them. This is called the T-Tauri wind, and due to it the gas and dust around those stars are expanding at hundreds or even thousands of kilometres per second.

These stars are believed to represent an evolutionary stage through which the Sun also passed while its thermonuclear reactions were stabilising. According to some researchers, notably A. G. W. Cameron and F. Perri in America, the density at the centre of the Sun *had* to be at least 100 times as high as it is today before it could ignite and start the T-Tauri wind blowing.

The wind may have lasted for a million years, acting as a 'new broom', sweeping the Solar System clean of gas and dust and leaving only the solid bodies that were in orbit. In fact only half as much material was left, altogether, in the system when the T-Tauri wind ceased. Even the inner planets were stripped of most of their primitive atmospheres – especially hydrogen and helium. The outer planets had already grown large enough and trapped sufficient material to hold on to a large part of their gases.

For several hundred million years the Solar System continued to settle down. The orbits of smaller bodies were continually perturbed by the gravitational effects of the major planets. Some tagged on as minor satellites, others were hurled out of the system completely by the 'sling-shot' effects of the giant planets, especially Jupiter (as used much later by Man with probes such as Voyager). Many more continued to bombard the planets and their satellites, leaving as evidence the craters we still see today.

## Final Stages

Even after the T-Tauri wind ceased, sunlight continued to influence small bodies such as asteroids or meteoroids up to a kilometre across. The radiation from the sunlit side of a rotating body can actually exert a thrust, which will cause the object to move in toward the Sun or away from it, depending upon the direction of rotation. It may then be captured by a planet when it approaches closely enough.

The analogy of the giant planets and their moons as miniature solar systems was even more close in the early stages. Jupiter, for instance, radiated about 1 per cent as much energy as the Sun for the first 10,000 years of its life, due to gravitational forces. But it could never attain the immense pressures and temperatures to make it ignite.

Next, we will take a detailed look at the enormous powerhouse on which the rest of the Solar System depends – the Sun.

■ 1 Protoplanets begin to form in the early Solar System, the inner ones collecting mainly rocky materials. The outer planets receive the bulk of the hydrogen and other gases, and their gravitational whirlpools draw in dust and smaller bodies, to become satellites. Meanwhile, the protosun is about to ignite and become a true star.
2 The Sun has ignited, and now the T-Tauri wind sweeps gale-like through the system, driving off the primitive atmospheres of the planets like comets' tails.
3 The Solar System is left relatively free of gas, dust and other debris. The Zodiacal Light – a residue of small particles – can be seen here surrounding the brightly shining Sun in the plane of the system.

# The Sun Today

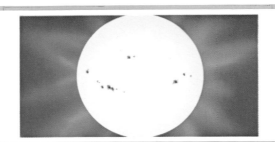

- Mean diameter: 1,392,000km.
- Equatorial rotation period: 25.4 days.
- Inclination of equator to ecliptic: 7°15'.
- Mean density (water = 1): 1.4.
- Rate of conversion, mass into energy = 4,000,000 tonnes per second.
- Temperature at centre: 14,000,000°C.
- Temperature at surface: 5,500°C.

*'If the Sun had been unattended by planets, its origin and evolution would have presented no difficulty.'*
Astronomy and Cosmogony SIR JAMES JEANS (1929)

To the inhabitants of planet Earth, the Sun appears ever-changing. A red, flattened globe sitting on the horizon at sunrise or sunset; a pale yellow disc sailing above hazy clouds; a bright, fuzzy glow surrounded by rainbow-coloured haloes; a blazing orb too bright to be looked at with the naked eye. Of course, we know that these are mere effects of Earth's thin film of atmosphere. The Sun itself shines steadily (with very minor variations), producing a flow of light and heat as it has done for billions of years and will continue to do so for billions more. The Sun is actually about halfway through its 10 billion-year life.

Even to the unaided eye, the Sun's surface (at sunset, say, or when veiled by thin cloud) sometimes appears to be freckled by dark spots. The ancient Chinese knew of these and called them 'flying birds'. To Aristotle and the Greeks, and to much of the Western World that inherited their learning, the Sun consisted of pure fire; and for hundreds of years no suggestion of a blemish would be countenanced, especially by the Church. This 'blind eye' attitude persisted even after Galileo, using the newly invented telescope, clearly saw spots which disappeared then reappeared.

Even 200 years ago the Sun was thought to be a molten body of glowing lava, and the spots were the tops of volcanoes or mountains protruding like islands in a sea. Then in 1774 the British astronomer Alexander Wilson pointed out that the spots must be craterlike depressions sloping to a dark interior. The German-British astronomer Sir William Herschel developed this idea, and from 1795 until well into the 19th century his view of the Sun was accepted.

According to him there were two layers of cloud: an upper, glowing layer with occasional gaps (the spots), and a lower one that protected a solid surface below from the heat of the top layer.

Around 1826 John Herschel in England and C. S. M. Pouillet in France attempted to determine the actual temperature of the Sun's surface. But the question was not resolved until 1879 when work by the Austrian physicists J. Stefan and L. Boltzmann gave a value close to 6,000°C, which is accepted today. At that time the actual source of energy was unknown. Helmholtz's contraction theory, proposed in 1871, was accepted for almost half a century. In fact, though this theory accounted well enough for the Sun's radiation *before* it ignited, but not after.

Another naked eye (if brief) method of observing some of the Sun's physical features is during a solar eclipse, when the Moon covers the Sun's disc, obscuring its glare, for up to seven minutes.

While this takes place the Sun's atmosphere – the chromosphere – and the pearly wings of its corona become clearly visible around the blacked-out Sun. If one is lucky the pink, feathery flame of a large prominence may be seen close to the edge of the disc. A telescope makes these plainly visible, though

---

'Pure fire is to be seen best in the sun, which is lit up fresh each morning and put out at night. It and the other heavenly bodies are just masses of pure fire ignited in a sort of basin in which they traverse the heavens, and this fire is kept up by the exhalations from the earth.'
Herakleitos (5th century BC)

---

great care is always needed when observing the Sun. The only safe method of doing so by telescope is to project its image on to a piece of white card, when sunspots and bright patches called faculae become easily identifiable. The main features are shown on the diagram opposite.

In order to find out what the Sun is made of, an instrument called a spectroscope is used, usually in conjunction with photography today. Although Sir Isaac Newton had split a beam of sunlight into its constituent 'rainbow' colours – its spectrum – with a glass prism as early as 1666, it was not until

the 1800s that the science of spectroscopy began. It was in 1802 that an English scientist, W. H. Wollaston, noticed dark lines in the spectrum of the Sun. A German physicist, J. von Fraunhofer, explained these in 1814 as being caused by different chemical elements, each of which *absorbs* its own characteristic colour if it is present in the incandescent light source. By 1817 he had identified hundreds of these 'Fraunhofer lines', assigning them letters – A, B, C and so on – which are still used. The spectroscope can also be used to find what gases and elements exist in the atmospheres of other planets or on distant stars.

Another instrument, the spectroheliograph, can enable us to photograph the Sun in the light emitted by any one specific gas at a particular atomic state caused by temperature and pressure. For instance, a well-known red colour identifies the hydrogen alpha line ($H\alpha$) to astronomers.

By this method, and others, we find that the outer portion of the Sun is composed of about 75 per cent hydrogen, nearly 25 per cent helium and a smattering of heavy elements. Once, the core had the same composition, but it is in this inner sphere, occupying a quarter of the total diameter, that 99 per cent of the Sun's energy is produced. As a result, after 4.5 billion years of nuclear fusion, there is more like 35 per cent hydrogen, 65 per cent helium in the core.

The most important fusion reaction producing all this energy is the proton-proton cycle – a three-part sequence, as shown on page 22. (There are several alternative sequences that take place).

Most of the photons released are gamma-ray and X-ray photons (meaning that they are at the high-energy end of the electromagnetic spectrum, beyond ultra-violet). Because of the immensely high temperature and density – 12 times that of lead – at the core, the gas there is completely ionised. All those free electrons and nuclei scatter and collide

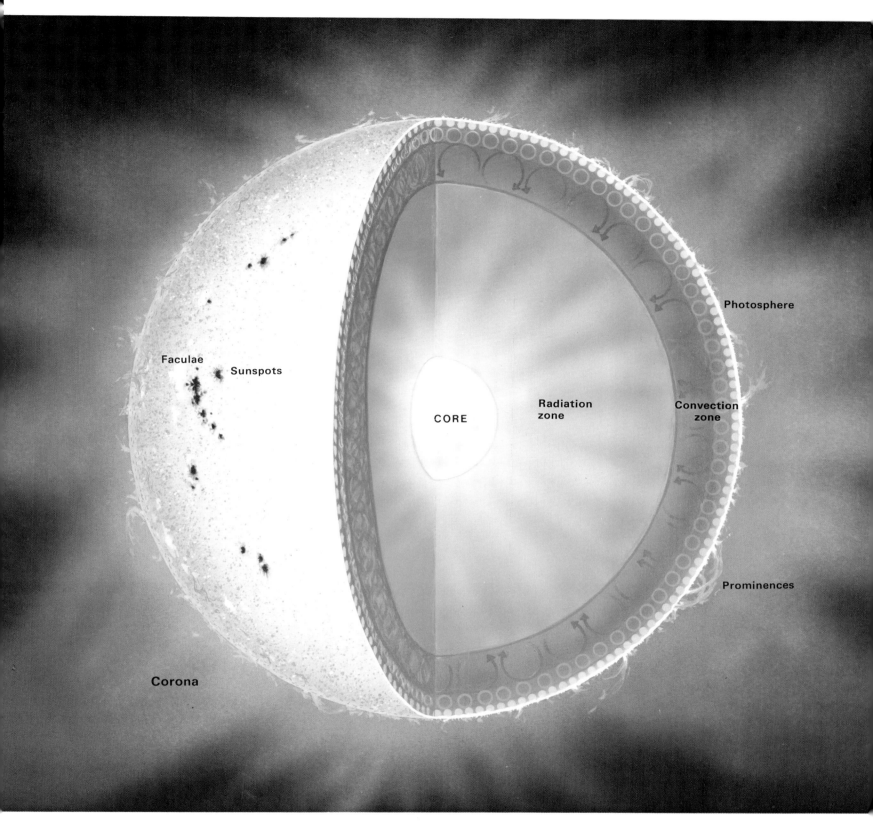

■ ABOVE: A cutaway of the Sun, showing its main physical
features. Sunspots are composed of two parts: the dark,
inner umbra and the outer, striated penumbra.

The corona varies in size and shape with the solar cycle.
At sunspot maximum it is twice as dense, and its
temperature is 20 per cent higher, than at minimum. It is
also more extensive at maximum, but then exhibits less
variety in shape. Note the 'polar plumes' near the Sun's
magnetic north and south poles – just one of the typical
characteristics of the corona's loops and streamers.

■ RIGHT: Before the invention of a special viewing device,
such sights could only be observed during a total eclipse
of the Sun (far right); the other illustration shows the more
common partial eclipse. But in 1930 the French astronomer
B. Lyot invented an instrument that covered the Sun's
disc, causing an 'artificial eclipse'. This enabled the
chromosphere and corona to be studied at almost any time,
with spectograph or any other instrument. N.B. The only
safe way to observe the Sun is to project its image onto a
white card.

$10^{-14}$   $10^{-13}$   $10^{-12}$   $10^{-11}$   $10^{-10}$   $10^{-9}$   $10^{-8}$   $10^{-7}$   $10^{-6}$   $10^{-5}$   $10^{-4}$

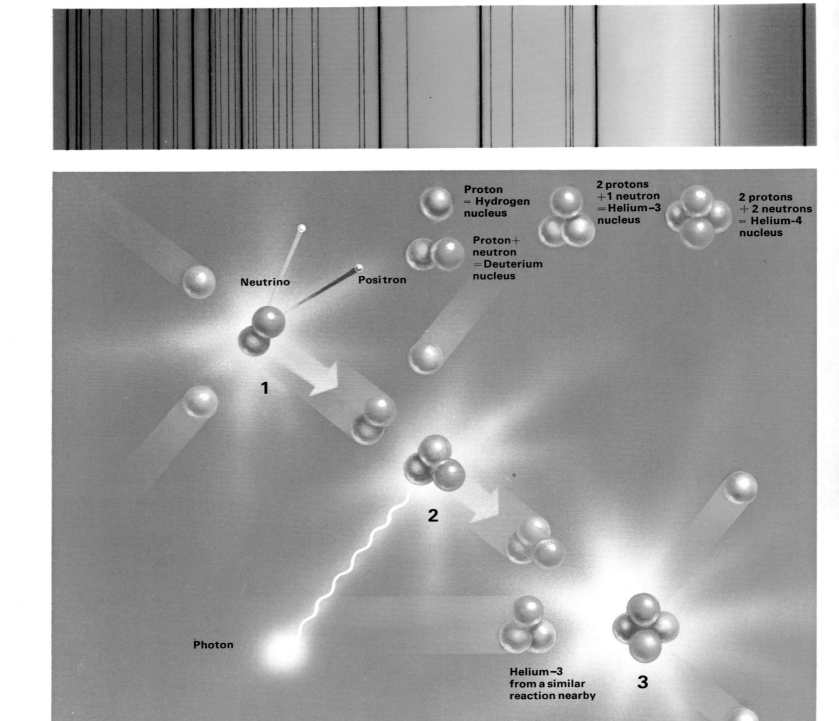

**Gamma rays**

**X-rays**

**Ultra-violet**   **Visible light**   **Infra-red**

■ TOP: The electromagnetic spectrum, showing the narrow band of visible light, and also the radio wavelengths which can pass through our shielding atmosphere (to be received by radio telescopes). ■ BELOW: the visible light spectrum of our Sun. The black absorption or Fraunhofer lines are shown, revealing the presence of hydrogen and various metals. The strong lines at the violet end are caused by potassium. In the early days of spectroscopy the lines caused by the highly ionised atoms of iron, nickel and chromium in the corona were thought to be emitted by a new element, which was named 'coronium'. ■ BOTTOM: The 'proton–proton cycle' which provides the main fusion reaction in the Sun's core. Only when the energy released reaches the photosphere does it become visible as sunlight.

Proton = Hydrogen nucleus

2 protons +1 neutron =Helium–3 nucleus

2 protons + 2 neutrons = Helium-4 nucleus

Proton+ neutron =Deuterium nucleus

Neutrino     Positron

**1**

**2**

Photon

Helium–3 from a similar reaction nearby

**3**

10⁻² 10⁻¹ 1 10 10² 10³ 10⁴ 10⁵ 10⁶ 10⁷ 10⁸ metres

# RADIO WAVES

FM radio       AM radio

'Radio window'

Television

Radar

Microwaves       Short-wave

with photons, which have to fight their way to the surface. They can take a million years to do so!

As the diagram on page 21 shows, the energy from the core travels most of the way by radiation. From about 150,000km below the surface the gases become even more opaque to radiation; the temperature has dropped sufficiently for atoms to be only partly ionised now, but their outer electrons absorb photons. Convection takes over at this stage and layers of huge convection cells, decreasing in size towards the surface, carry the energy outward.

The final layer of these cells forms the photosphere – the opaque but glowing visible surface – which in close-up resembles cumulus clouds as seen from above in our own atmosphere. In the case of the Sun, though, it presents an overall mottled surface, an effect known as granulation. Each granule is from 1,000 to 2,000km across.

In a telescopic view of the photosphere, the edge or limb of the Sun appears darker than the middle. This is because at the centre we look 'down' into the hottest, brightest layers of the 500-km deep photosphere, whereas at the edge we look at an oblique angle through relatively cooler layers.

Because of its temperature, the light emitted by the photosphere is yellowish; if it were cooler it would be more red, if hotter, bluish. All stars are categorised according to their temperature-colour range. Having left the photosphere, energy then again travels by radiation through space to heat and light the planets and other bodies.

Not only visible light ('visible', that is, when it strikes something which reflects it) and radiant heat (infra-red) are emitted by the Sun, of course. Invisible radiation, from gamma-rays to radio waves also bathe the Solar System. For convenience, we think of the shorter wavelengths as particles – photons, or 'bundles of energy' – and the longer wavelengths as waves. Visible light can behave as either, on occasion. There is also the Sun's magnetic field.

But in addition the Sun emits vast quantities of strange particles called neutrinos.

These are produced in the core as a by-product of the fusion reactions, but unlike the photons are not obstructed by the nuclei and free electrons, and pass straight out into space at the speed of light. If we could examine neutrinos we should learn more about the workings of the Sun's core. However, these elusive particles appear to have no electrical charge and negligible mass, so rarely interact with other particles, even in large aggregations of matter such as the Earth. With these peculiar properties, how can we hope to detect them?

It happens that when a neutrino meets an atom of chlorine-37 it produces an atom of the rare gas argon-37. So in 1968 American chemist Raymond Davis, Jr., built a very large tank and filled it with a dry-cleaning liquid, perchloroethylene ($C_2Cl_4$), which contains a lot of chlorine.

## The Neutron Trap

To shield it from other forms of radiation he located his tank at the bottom of the very deep Homestake gold mine, near the town of Lead in South Dakota. The tank was flushed out at regular intervals and special equipment counted the number of newly formed argon atoms. Unfortunately, the number of these proved to be far too low – about one third of the 'neutrino flux' calculated to be released. The equipment was re-checked and found to be sound, and the astronomical world was thrown into a discussion which has lasted for years, and continues (especially in the light of recent evidence).

One explanation was that the temperature of the Sun's core is lower than expected. Other explanations involve very strong magnetic fields, a rapidly rotating core, and a smaller proportion of heavy elements in the core than expected. Thus the Sun's interior would be more transparent to radiation, so a lower core temperature (due to slower

reactions) would release fewer neutrinos. However, the amount of heavy elements found spectroscopically at the surface is many times too high to confirm this. A recent theory suggests that a higher proportion at the surface could be due to 'contamination' by interstellar dust particles since the Sun was formed. There are arguments against this, but the whole episode illustrates the fact that our knowledge of our parent star is far from complete.

The presence of the photosphere explains why the Sun's disc has a sharp edge instead of being fuzzy as one might expect of a ball of hot gas.

As for those black-seeming sunspots, they appear dark only because they are some 2,000K cooler than the surrounding photosphere. If the Sun were one gigantic sunspot we would notice only a small difference in the strength of sunlight (visually), while one large spot, alone against the blackness of space, would give as much light as the full Moon.

As Wilson had surmised (page 20), sunspots are indeed depressions. The largest recorded sunspot, which occurred in 1858, was almost 20 times as wide as the Earth! A more usual size would be 10,000km across; the smallest, called pores, last only hours; the largest last for weeks, passing out of sight as the Sun rotates and then reappearing on the opposite limb.

Spots often occur in pairs or groups, and are strongly magnetic in nature. The pairs are always of opposite polarities – north or south – and align themselves with the Sun's equator. Strong magnetic fields affect the lines in a spectrum and even enable us to measure the strength of the field in individual spots, which proves to be thousands of times greater than Earth's own magnetic field.

The Sun's general surface magnetic field is not much stronger than Earth's – 1 unit of magnetic induction, or gauss, compared with Earth's 0.6 gauss. Yet the field around a sunspot may be as strong as 4,000 gauss. The reason for this is that, being gaseous, the Sun does not rotate as a rigid body. Whereas a point on the equator takes 26 days to make one rotation (as seen by a 'static' observer, *not* from Earth, which itself moves around the Sun), a point near the pole takes 37 days.

As we saw, much of the matter in the Sun is ionised. Such electrically charged gas, or plasma, is a good conductor of electricity. It is therefore strongly influenced by a magnetic field and normally tries to follow the direction of its 'lines of force'. A magnetic field can, however, become 'locked' into the plasma and then, because of the differential rotation of the photosphere between equator and poles, and the fact that the core probably rotates at a different rate altogether, the field becomes twisted around as shown at the top of page 25. In doing so its strength is progressively increased.

However, the pattern is not as simple and regular as shown here. The field is violently contorted into vortices and loops, particularly towards the end of the cycle. Where a magnetic loop projects above the surface its points of exit and re-entry are marked by a pair of sunspots, of opposite polarity. The field is also distorted by convection currents and other turbulence within the Sun.

The whole cycle takes 22 years to complete. As one cycle ends, a few spots may be seen near the Sun's equator. Then spots begin to appear at higher latitudes – about 40°N or S – until, at sunspot maximum, many groups are found around 20°N or S, gradually approaching the equator, which they reach at the end of 11 years with a new minimum.

When the new cycle starts, something interesting happens. The spots which appear, again around 40°, have reversed their polarity! If the leading or westerly spot of each pair in the northern hemisphere was 'north' magnetically (and vice versa in the southern) it now becomes 'south' while its counterparts in the southern hemisphere become 'north'. It thus takes 22 years to return to the original polarities. In addition, the *whole* of the Sun's magnetic field reverses about every 11 years.

In addition to the granules which mottle the photosphere there is a larger-scale effect termed supergranulation. Supergranules are some 30,000km across, and gas flows outward towards their edges – an effect resembling soup boiling in a pan. Then there are short, dark lines seen against the Sun's bright surface, called fibrils; when seen at the limb, against a dark background, they resemble gas flames or jets and are then known as spicules. Each lasts only a few minutes. Small though these look, they

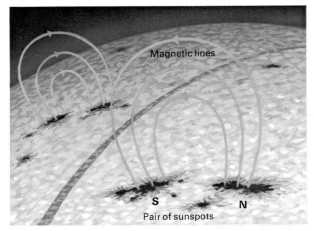

Magnetic lines

S    N
Pair of sunspots

■ LEFT: Pairs of sunspots appear to show where magnetic loops from the interior have burst through the surface, giving rise to opposite polarities. The magnetic field is probably twisted and distorted by the circulating gas in the convection zones. ■ BELOW: solar prominences. Here, too, the magnetic lines of force are clearly revealed by the material as it condenses along them. The longest-lasting prominences are called 'quiescent', and can build into massive structures resembling arches or forests with interlinked stems. 'Eruptive' prominences may spurt almost a million kilometres into space.

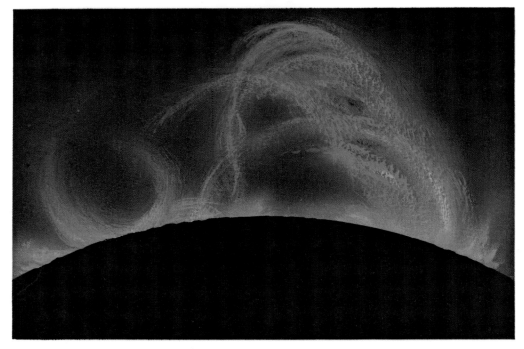

actually rise to heights of 10,000km. It is easy to forget the gigantic scale of the Sun.

It is in the areas of high magnetic activity around sunspots that other solar phenomena are likely to be seen. Before a spot appears, the chromosphere (the pinkish 2,500km-thick frothy layer into which the spicules rise, immediately above the photosphere) may be violently disturbed, an area known as a plage; the concentrated magnetic field in the area probably has an intense heating effect locally. The photosphere itself can also brighten, forming the light patches called faculae.

## Solar Fireworks

Another form of disturbance on the solar surface is due to flares. Again usually seen in an active region, these are brilliant, starlike flashes often lasting only a few seconds, and no longer than an hour. Flares actually release a high spectrum of energy, from X-rays and ultra-violet to radio waves, in a brief burst. When these reach the Earth their effects are felt in various ways, including interference and fading on short-wave radio.

The most beautiful displays, however,

are the rosy-hued prominences. Contrary to the effect often given by photographs, prominences do not rise like flames out of the photosphere and arc into space. Except in the case of eruptive prominences, in which matter does appear to be expelled from the surface with considerable violence, the prominence materialises as if by magic out of the corona and slowly falls towards the photosphere. When prominences are seen against the bright surface of the Sun they appear as dark streaks called filaments.

The corona itself consists of hot, rarefied gas extending for millions of kilometres, eventually merging into space. Because it could only be viewed during an eclipse (it is normally 'swamped' by light scattered by our atmosphere) it was often said to be merely an optical illusion.

Where the chromosphere blends into the corona the temperature rises rapidly to a million K or more. It was a fact that long puzzled scientists, because the temperature of the photosphere is measured only in thousands of degrees. It seems to be due to supersonic shock-waves from the convection currents bursting in the photosphere.

■ ABOVE: A single magnetic line will be distorted as the photosphere rotates, differentially, from 26 days at the equator to 37 days at the poles. It is also forced to move with the ionised gas or plasma, which has a high electrical conductivity, and becomes wrapped around the Sun. Convection and other turbulence within the Sun distort the magnetic field still further, and by the end of 11 years it becomes hopelessly tangled and so neutralises itself. A new field then forms, and the cycle is repeated.

■ BELOW: Earth's magnetosphere. The solar wind is distorted by a planet's magnetic field, encasing it in a sort of protective force shield. The 'bow shock' is resilient and 'springy', absorbing the varying solar wind. The Van Allen radiation belts were detected by the first Explorer satellite in 1958 and named after US scientist Dr. J. A. Van Allen, who first mapped their shape and distribution.

Magnetosphere  Magnetopause  Plasmapause  Ring currents  Van Allen belts  Bow shock wave

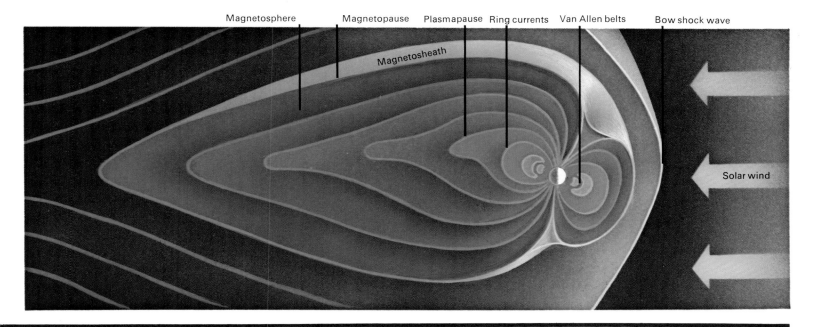

Magnetosheath

Solar wind

Energy released from flares and magnetic effects probably also add to the frictional effects of dissipating shockwaves in heating this transition zone. It should be remembered, though, that the coronal gas is thin, so there is little actual 'heat' contained in it – the atoms are widely separated. Highly ionised atoms of iron, nickel, and chromium have been identified in the corona.

The corona displays a whole range of structural details – loops and streamers especially, which indicate the magnetic flux controlling them. Of particular significance are the polar plumes which, as their name implies, spring from the magnetic north and south poles of the Sun. At its outer limits the gas of the corona merges imperceptibly with the Zodiacal Light, a swarm of dust particles orbiting the Sun and reflecting its light, in the plane of the Earth's orbit, known as the ecliptic. The Zodiacal Light can be observed occasionally as a faint cone of light after sunset or before dawn (when it is sometimes called the 'false dawn').

But, as we have seen, the Sun's influence pervades the whole Solar System. The high-energy particles released by flares and sunspots stream outward through the system, together with the plasma of the corona itself as it expands into space. This out-rushing gas is known as the solar wind.

By the time it reaches the Earth – taking 2–5 days to make the trip – it may still be travelling at 600km/s or more, but its temperature has fallen from the millions of degrees at the corona to perhaps 200,000K. Because it is so rarefied, it has no measurable heating effect.

## Solar Influences

It was not until the advent of solar observations from the US orbiting space laboratory, Skylab, that it was realised that the solar wind emanating from the Sun's polar regions travels twice as fast as that from the equatorial regions, and that the former originates from 'coronal holes' over the poles, even though there is no visible activity in these regions.

Apart from interference with radio caused by solar X-rays – the strength of which may increase 100 times during a flare – the solar wind has many observable side-effects. As we saw, ionised matter is forced to follow magnetic lines, so when it encounters a planet that has a magnetic field it forms a sort of shell – the magnetosphere. In the case of Earth the ionised particles are trapped in the upper atmosphere, creating the ionosphere. This reflects longer radio waves and enables radio transmissions to be literally bounced back to the ground to reach receivers all over the world. But their most beautiful byproducts are the aurorae. From countries near to our poles (where electrons are able to spiral in, collecting in areas known as the auroral ovals), the many-coloured, swirling curtains of the aurora have long entranced and awed all privileged to see them.

The Sun is also itself a gigantic radio transmitter. Wavelengths vary from very short in the lower chromosphere to longer waves from the corona, due to the decreasing density of free electrons with distance (these absorb longer wavelengths). There are also bursts of radio waves, of at least five types.

Recent observations suggest that our star may not be as stable and unchanging as was once thought. It apparently 'pulses' – admittedly by a mere 10km – every 160 minutes. It seems, then, that in spite of the enormous strides in our knowledge of the workings of the Sun, solar scientists still have many mysteries to unravel.

# Earth Evolves

- Mean diameter: 12,750km.
- Escape velocity: 11.2km/s.
- Rotation period: 24 hours.
- Sidereal (orbital) period: 365.25 days.
- Mean distance from Sun: 149,600,000 km.
- Inclination of equator to orbit: 23.45°.
- Mean density: 5.5.
- Present atmosphere: Nitrogen/oxygen: ratio 4.1.

*'Nowhere in the solar system but on our own earth can we find an oxidising atmosphere. Nowhere else is there free oxygen. Why?'*

Is Anyone There ISAAC ASIMOV (1967)

During the formation of the Solar System, the protoplanet that was to become Earth condensed and slowly grew by accretion, sweeping up vast quantities of particles from the nebula surrounding the protosun, which finally ignited to become the Sun.

By some 4.5 billion years ago the primeval Earth was being bombarded by planetesimals. Although relatively cold themselves, these rapidly heated the planet as the gravitational potential and kinetic energy of their falling was converted into heat energy upon impact. Additional heat was generated by compression, as the growing mass on the outside squeezed the interior, and by the radioactive decay, or disintegration, of certain elements such as uranium, thorium and potassium-40 within the Earth's interior. The latter is a slow process, since the heat produced by radioactivity is due only to the energy of the movement of particles emitted (consisting of helium nuclei and electrons) being absorbed by the material through which they pass. But it is a steady process, and one which still continues. Because rock is a poor conductor of heat, the heat generated by these elements is 'trapped' in it.

The interior of Earth consisted at this time of a uniform mixture of silicon compounds with iron and magnesium (mainly combined with oxygen in the form of oxides) plus smaller quantities of all the elements that make up everything we know and use today. About a billion years (or less) after Earth formed, the temperature some 400–800km below the surface reached the melting point of iron, which, being heavier than the materials around it, sank towards the centre. Here it formed a large liquid core; and even the gravitational energy of its 'falling' released more heat – eventually rising to some 2,300K and melting a large portion of the Earth's interior.

As the iron and other heavier metals sank, lighter materials floated upward forming a mantle around the core and an outer crust. Some lighter metals, such as aluminium, potassium and sodium, accompanied the silicates and other non-metallic materials into the crust. The heavy radioactive metals uranium and thorium also rose with them, due to the crystalline compounds into which they had formed and which could not exist in the high pressures near the core.

The crust itself began to form into layers. At the top, containing the light metals, was silicon-rich granite which had in turn solidified and crystallised from the molten basalt and gabbro of the mantle. The crust is very thin: at

---

'Had I been present at the birth of this planet I would probably not have believed on the word of an archangel that the blazing mass, the incandescent whirlpool there before our eyes at a temperature of 50 million degrees would presently set about the establishment of empires and civilisations . . .'
*The Human Situation*
W. Macneile Dixon (1937)

---

most 70km under mountain ranges, but normally 35km under land and often only 5km under oceans.

The core, too, is probably composed of two parts. While the outer core (which is slightly over half the diameter of the whole Earth) is undoubtedly liquid, the inner core – itself as big as the Moon – has been compressed into a solid state by the immense pressure. The temperature at the centre may be as high as 4,500K.

Dr. William Gilbert, physician to Queen Elizabeth I, had shown that Earth's magnetic field was like that from a bar magnet. But the discovery that the centre of the Earth was molten caused some consternation; permanent magnetism is *lost* at high temperatures – well below the melting point of iron! Scientists got around this by assuming that the liquid core is a conductor of

electricity – iron – and contains 'loops' of current which, as the core rotates, act like a great dynamo to produce the field which extends several Earth-diameters into space and diverts the solar wind to form the magnetosphere.

Like the Sun, Earth's field seems to have reversed its polarity, but unlike the Sun's 11-year cycle, Earth's takes 200,000 years (even then, not regularly). It is a fact recorded in rocks and in cores taken from ocean beds. Indeed, it seems that eight different species of a microscopic marine organism, *radiolaria,* became extinct exactly at the time of one reversal, after existing successfully for millions of years. It is widely believed that creatures such as fish, bees and birds use the magnetic field as an orientating device. (The next reversal is greatly overdue.)

Some 4 billion years ago – the age of the oldest rocks found today – there were no continents as such. Below the crust, as the material in the mantle became molten (heat being able to escape only slowly to the surface by conduction) a new method of heat-transfer came into action: convection.

Authorities are still divided upon the exact mechanism; whether it extends throughout the entire mantle or is limited to 'cells' nearer the surface, as in the Sun, and so on. But the existence of such a continuous process, carrying hotter and lighter (because less dense, since hot substances expand) material from below, while cooler matter near the surface sinks, was crucial to the formation of the continents as we know them today. Its first effect, though, was to aid rapid cooling of the early Earth and the separating out of chemical elements and compounds (minerals); those of lower density tending to collect at the surface.

Among these were radioactive minerals, which, having 'settled' in the crust, continued to produce heat. Being near the surface, this could now escape

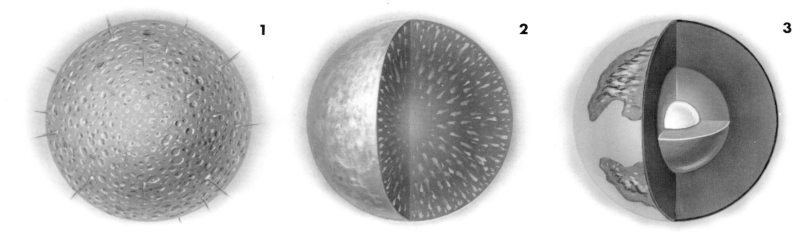

■ ABOVE: Three stages in the evolution of Earth.
**1.** Millions of planetesimals plummet through the hydrogen-rich early atmosphere, accreted from the nebula. The surface is pocked by craters, like today's Moon.
**2.** Several tens of millions of years later, the effect of impact heating and internal radioactive decay, plus compression due to gravity, causes differentiation: metals and heavy materials sink toward the core while lighter materials float upward to form a crust. The solar wind has stripped the hydrogen atmosphere, to be replaced by a new atmosphere caused by 'outgassing' from within the Earth.
**3.** Over 3 billion years ago, the water in this primordial atmosphere has condensed, and the first continents have arisen, ravaged by volcanic activity. The core, mantle and crust have settled into almost their modern conformation, but crustal plates, subduction and sea-floor spreading have not yet appeared.

■ BELOW: A cutaway of today's Earth (*see text*). Geologists discover what lies below the surface by setting off explosions and recording with a seismograph the waves that travel through rock of different densities at varying rates. The boundary between the crust and mantle is called the 'Mohorovičić discontinuity' (M-discontinuity or Moho for short), and here the waves' velocity increases sharply. There is also an increase in velocity at the 'transition zone', whose depth is between 350 and 700 km, but this contains several sub-zones, each with a different density.

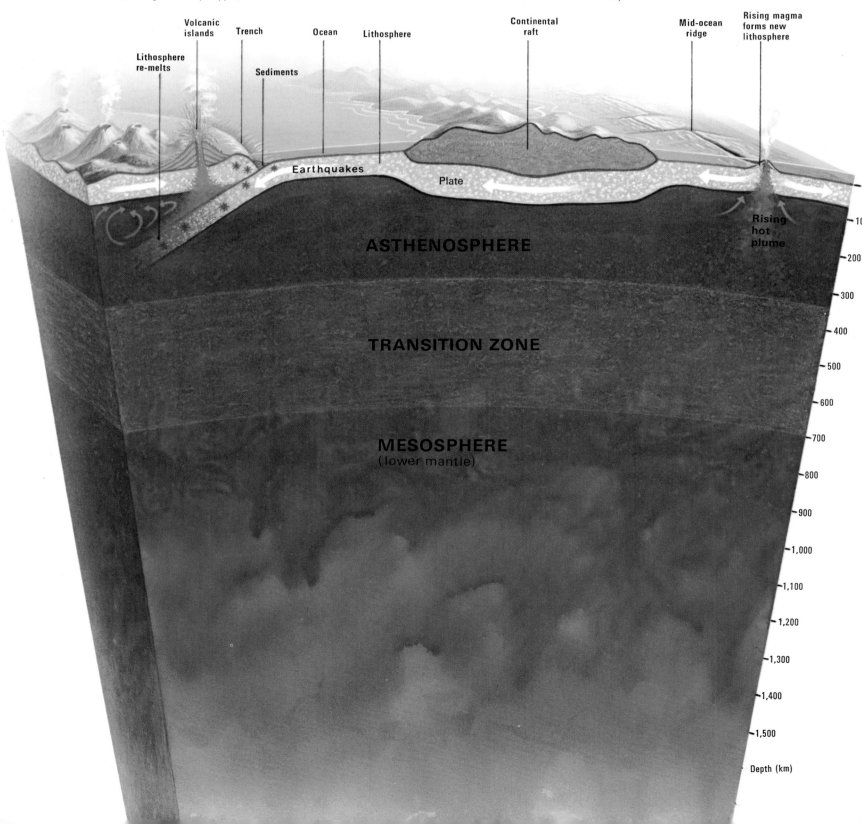

more easily. And, because the rate of radioactive decay is fixed and unaffected by chemical changes, heat or pressure, we are enabled to date the rocks exactly, 'simply' by finding the ratio of the original element to the new element(s) into which it has decayed.

The earliest continents were merely an outer crust formed by lava welling up from below, solidifying, re-melting, and so on. With the lava came gases, which poured out of the Earth in clouds – a process called outgassing. Many of the minerals contained water in chemical combinations, and this was released when the rocks in which the minerals occurred were melted. So the initial atmosphere contained water vapour, along with carbon monoxide and dioxide, nitrogen, hydrogen (which soon escaped, being light), and other gases, some in combination. Eventually the surface cooled sufficiently for water to condense and, more important, for it to remain liquid upon reaching the ground instead of instantly vaporizing. Rain must have fallen in unimaginable torrents.

On the surface, impact cratering continued, though at a much reduced rate, breaking through the crust in places, heaping it up in others, and perhaps even forming the first ocean basins. But the earliest craters had already been wiped away by the continuous re-melting process. Meanwhile, down below, the mantle had cooled enough to partially solidify. This did not stop the convection process: convection can take place in solids as well as liquids and gases if the conditions are suitable, and even 'solid rock' will flow and rise as its density is reduced by heating from below. It is convection that provides the driving force for the process called continental drift.

The unifying theory by which this is linked with mountain-building, the distribution of volcanoes and earthquakes, ocean trenches and ridges, etc., is known as plate tectonics.

Anyone interested in maps must have noticed that the outline shapes of continents – eastern South America and western Africa, for instance – seem to suggest that they were once joined. Certainly Francis Bacon did in 1620, but it was not until 1912 that a German geophysicist, Alfred L. Wegener, first presented the idea of continental drift as a serious scientific theory, at a lecture. As it happens, two Americans, F. B. Taylor and H. B. Baker, had also independently suggested similar explanations a few years earlier, but it was Wegener who did so in greatest detail and produced the most evidence. The immediate reaction of the scientific community was cool, to say the least. It took ten years for his hypothesis to be generally approved in his own country, and 50 to be universally accepted and taught.

The drawing below shows how our present-day continents could once have formed a giant super-continent, which Wegener called *Pangaea* (from the Greek: 'all land'). He found correlations in geological features on opposing coastlines, and even in fossils in coal deposits. It was only in his proposed mechanism for the movement of continents that his theories were weak, and these were attacked by Harold Jeffreys in England. Reaction became even stronger after Wegener's death in 1930; it was the acceptance of the existence of convection currents in the mantle, as theorised by two geologists, A. L. du Toit, a South African, and A. Holmes in Britain, that won his theories recognition.

## Plate Tectonics

As the illustration opposite shows, the continents ride like rafts, partly submerged in the solid lithosphere, which in turn 'floats' on the partially molten asthenosphere. Each of the ten or so plates into which the lithosphere is broken moves in a specific direction at its own speed. Where two meet head-on at a convergent junction, mountains are thrust up, volcanoes erupt and earthquakes are probable, as one plate rides up on the other, which sinks and re-melts, providing an underground reservoir of magma; this erupts at the surface as lava. The area where a plate dives under another and melts is known as a subduction zone. Where this occurs under an ocean a deep-sea trench is formed.

Where plates move apart, at a divergent junction, the crack is filled by molten material from below the lithosphere. As this solidifies it causes the plates to grow, continuously forming new oceanic crust, a process known as sea-floor spreading. It creates a ridge which travels through the major ocean basins – the Mid-Atlantic Ridge, for instance, where North and South American plates and the Eurasian and African plates separate. Again, volcanoes and earthquakes abound in these areas.

The continents we know today moved apart comparatively recently – during the last 5 per cent of Earth's geological time-scale. The two major land-masses, the one in the northern hemisphere called Laurasia and the southern Gondwanaland, began to split apart some 300 million years ago, and virtually all the present oceans have been created during the last 100–200 million years. (Without the destruction of the plates at subduction zones, the Earth would gradually increase in size.)

When the hot molten basalt emerges from the sea-floor ridge and cools below what is known as the Curie point, it becomes magnetised in the direction of the Earth's magnetic field *at that time*. Magnetometer surveys of the oceanic crust have confirmed that the parallel strips along the borders of the ridges do indeed have the prevailing polarity

■ The first 'supercontinent' – Pangaea. The fit of today's continents is even better if one uses the edges of their continental shelves rather than their coastlines. The oldest mountain belts – formed over 260 million years ago – are shown in orange. These were caused by the collisions of even earlier sections of continent. In the Mesozoic era two continents, Laurasia and Gondwanaland, were separated by the Tethys Sea, but may have been connected by a land bridge in the west thus allowing various species of reptile to migrate from one continent to another.

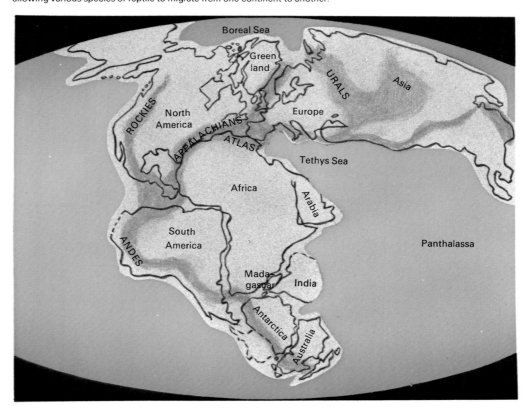

'frozen' into them. It is precisely these prevailing polarities, the reversals in Earth's magnetic field, that have enabled relatively accurate dating, and the record of these in the oceanic crust has shown that the sea-floor spreading proceeds at the rate of from two to 18cm per year.

So far we have looked only at the rocks that make up the surface of our planet. But there is something that exists on the surface that makes Earth unique in our Solar System: *life* – or the products or remains of it. A few hundred million years after water first condensed and formed the first large pools and shallow seas, primitive organic compounds were somehow synthesised.

The credit for originating the theory which is generally accepted today usually goes to A. I. Oparin, a Russian biochemist. In 1924 he suggested that the early atmosphere would have contained no free oxygen ($O_2$), but much hydrogen, mostly combined with nitrogen as ammonia ($NH_3$) or carbon as methane ($CH_4$); water ($H_2O$) is itself composed of hydrogen and oxygen. (Much of the carbon dioxide ($CO_2$) had combined with calcium and other substances to form carbonate rocks.)

Some form of energy, Oparin said, must have created organic molecules from these chemicals – molecules similar to those making up amino acids, those basic 'building blocks' which string together to form proteins. The early atmosphere contained no free oxygen to form an ozone layer ($O_3$) at high altitude (as there is today) to absorb the shorter, cell-damaging ultra-violet rays from sunlight. Strangely enough, it was the full blast of these that, together with intense lightning flashes in the cloud-laden atmosphere, provided the necessary energy to turn the early oceans into a 'soup' of cell-like organic molecules.

It was 30 years before Stanley L. Miller, in America, confirmed experimentally, in an apparatus consisting of a glass container full of methane, ammonia and hydrogen into which water vapour was introduced, that strong electrical discharges could indeed produce amino acids.

It was not known at the time of Oparin's initial hypothesis that the substance containing the 'genetic code' to reproduce and pass on inherited characteristics, ie 'evolve', was DNA (deoxyribonucleic acid). It is of DNA that our genes are made. Two biochemists, J. Watson and F. Crick, solved the secret of its complicated, double-helix structure in 1953. When cells divide each retains a strand of DNA, ensuring that it resembles its 'parent' but enabling it to evolve and change in response to external physical conditions or other stimuli (such as radiation). When the sequence of DNA molecules is different for any reason – whether beneficial or harmful – the result is a mutation.

It is a long, long step from those first living cells in the primordial soup to large complex creatures such as a whale, a man or even a cockroach, yet all are composed of aggregations of simple cells, of specialised types. The first cells (fossils of which have actually been found, 3.2 billion years old) were similar to modern bacteria, and lived off the soup itself. The most important mutations to follow were those that learned how to make their own food (sugars) using the energy of sunlight, together with water and carbon dioxide: the process of photosynthesis.

## Earth's Green Carpet

The cells that first did this were the blue-green algae, and their by-product was of supreme importance to Man and to all life on Earth today. They gave off oxygen. 'Blue-greens', as they are now generally known because they are more primitive than true algae, are still common today in the tropics. In Australia there are flourishing colonies of these *stromatolites* – and their billion-year-old ancestors are also found in the limestone deposits which they slowly form by their secretions. Like all plant-life today, they extracted hydrogen from water (earlier bacteria obtained it from gases such as hydrogen sulphide around volcanic vents, as some still do, but these were limited to such areas). The oxygen they gave off gradually changed the composition of the atmosphere; a remarkable reaction made possible by green pigment chlorophyll.

At this time, palaeomagnetic evidence shows, a major continent occupied the region which is now the South Pole, and the first known Ice Age may have begun. For a tremendously long time in Earth's history – 3–4 billion years – the only life consisted of single cells. This is known as the Proterozoic Era. Then, within the next hundred million years or so, perhaps because by then the oxygen in the atmosphere had built up to a certain critical level, there was an explosion of life, in an incredible variety of forms.

One important development was the grouping of cells which 'co-operate' and work together. For instance, cells that have cilia or flagella (hair-like filaments or longer whip-like tails which propel them around rather randomly) may group together, their flagella beating in unison to drive them in a desired direction.

At least equally important was the method of reproduction. Instead of one cell merely splitting, each 'twin' receiving a duplicate set of genes, there came a time when two different types of cell were produced; one large and not capable of much movement, the other small and propelled by a flagellum. By allowing combinations of different genes from two parents – one in the egg, the other in the sperm – the arrival of sex at last really allowed evolution to

| PRE-CAMBRIAN | PALAEOZOIC | | | | | | |
|---|---|---|---|---|---|---|---|
| | CAMBRIAN | ORDOVICIAN | SILURIAN | DEVONIAN | CARBONIFEROUS Lower | Uppe |
| PLANT LIFE: Algae, Lichens, Fungi | Seaweeds | Water plants | First land plants | Horsetails, Ferns, Club mosses | Swamp plants | First coni |
| ANIMAL LIFE: Single-celled animals, Bacteria, First shelled invertebrates | Brachiopods, Corals, Trilobites, Molluscs, Jellyfish | Jawless armoured fish, First nautiloids | True fish, First fish with jaws | Spiders, Scorpions, First fish, First amphibians | First insects | |
| -600 million years | -500 | | -400 | | | |

get under way.

Obviously there is not room here to catalogue all the varieties of plants and animals that evolved from then on. The stratigraphical diagram along the bottom of these pages charts some of the landmarks along the way (though it must be remembered that there was often considerable overlapping). The geological time-scale is divided broadly into eras: the earliest, in which the first cells appeared, is the Archaeozoic; then followed the Proterozoic in which oxygen-producing bacteria began enriching the atmosphere, sexual reproduction arrived and multicellular life-forms developed. From then on (in only the last 14 per cent of our planet's history) we can divide the eras into periods measured in hundreds or even tens of millions of years.

For a very long time organisms were dependent upon water for support, and it is often difficult to classify them as 'animal' or 'plant'. There were sponges, which appeared less than a billion years ago; then came forms of jellyfish and soft, then rocky, corals – tiny animal polyps that work together and secrete limestone 'skeletons' from their bases which form gigantic colonies. But these had to be anchored underwater; the most successful shape arrived at by evolution was the tube, with an opening at each end, through which nutrient could pass whilst being processed and absorbed. Stripped of all embellishments, this is all that animals, including humans, consist of today!

It is difficult for fossils of the first soft-bodied creatures to be preserved, though jellyfish have left their faint imprint – again in Australia, baked into wet sand that has become rock. Then, quite suddenly – no-one knows exactly why – just at the end of the Precambrian period some 600 million years ago, whole groups of organisms began to secrete hard shells, and so left records in the rocks. So we find the first trilobites and other shelled invertebrates (without backbone). Vertebrates came later, at the end of the Ordovician. But both backbones and shells serve as points of attachment for muscles, so were vital developments.

For their part, plants developed a cellular structure that provided rigidity by containing water *inside* their stems, and so evolved from primitive, ground-hugging mosses into club-mosses, ferns and horsetails. Varieties of these survive almost unchanged today.

The shells of those primeval aquatic creatures played a vital role in the evolution of Earth as a planet, though. As they died and fell to the sea beds in their untold billions the silica and calcium carbonate in their shells ended up as massive deposits which were in time to become rock – limestone, chert, etc. They are also the main source of phosphates.

A balance was established between the oxygen-producing vegetation and oxygen-breathing animals. (Such a fund of oxygen has been built up that it has been calculated that even if all photosynthesis stopped today the supply of oxygen in the atmosphere would last for over 2,000 years at the present rate of consumption by all users.) The formation of the ozone layer began to protect the land areas from ultra-violet radiation.

## Continental Thrust

Meanwhile, the early continents were in motion. An early collision towards the end of the Palaeozoic Era, 260 million years ago, between North America and the Africa–Europe plate probably not only wiped out vast populations of invertebrates living in the shallow waters of their continental shelves, but also created part of the Appalachian mountain range. Pangaea began to break apart in the early Triassic, at the beginning of the Mesozoic Era, with a split between Africa and Antarctica.

At that time the age of reptiles was under way, and the early dinosaurs had appeared. Flowering plants decorated the land, along with grasses and deciduous trees. The splitting of the continents eventually prevented the movement of land animals and isolated them on one continent or another, where they began to follow their own paths of evolution. This may not have occurred until the end of the Cretaceous, by which time mammals were making their takeover bid for the land, and the age of reptiles was almost over.

Erosion by wind, rain and extremes of heat and cold sculptured the early mountains, and the sediments from this process were carried down in rivers to the widening oceans. The layers of sediment laid down were often so thick that they covered the remains of marsh plants and forests (particularly in the Carboniferous period). The sediment hardened into rock, eventually kilometres thick, and under this great pressure the Coal Measures were formed. Similarly, but later – mostly in the Jurassic – microscopic plants called *phytoplankton* and the tiny animals that lived on them in the warm seas were also covered by layers of silt that hardened into sandstone and shale, forming oil and gas.

So, to the original igneous rocks (basalt, gabbro, obsidian, granite, etc.) produced from magmas beneath the lithosphere were added sedimentary rocks composed of rock fragments compressed and hardened, or of organic sedimentary rocks (including chalk) made up of the shells of sea-creatures, from single-celled diatoms and foraminifera to the much larger brachiopods and ammonites, whose fossils are often found in them.

Any of these may be transformed into metamorphic rocks by the later action

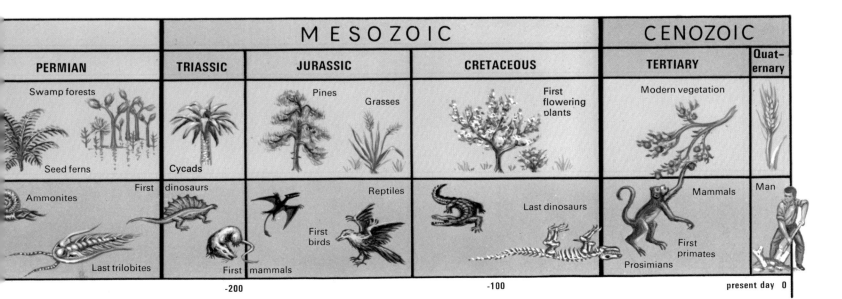

| | | M E S O Z O I C | | | | C E N O Z O I C | |
|---|---|---|---|---|---|---|---|
| PERMIAN | TRIASSIC | JURASSIC | | CRETACEOUS | | TERTIARY | Quaternary |
| Swamp forests | | Pines | Grasses | | First flowering plants | Modern vegetation | |
| Seed ferns | Cycads | | | | | | |
| Ammonites | First dinosaurs | | Reptiles | | | | Mammals | Man |
| | | First birds | | | Last dinosaurs | | First primates |
| Last trilobites | First mammals | | | | | Prosimians | |
| -200 | | | | -100 | | present day 0 | |

of heat or pressure or by being permeated by gases or fluids, usually resulting in a changed crystalline structure. Thus shale becomes slate or schist; limestone, marble; granite, gneiss; and sandstone, quartzite.

As layers of sedimentary rock are pushed together and squeezed by plate movements or other processes, they can be forced upward into complicated folds, which when weathered away may reveal fossils on a mountain top that was once an ocean floor. These are known as deformation structures – a term that includes any of the many types of folding and faulting that can occur in the upper layers of the Earth's crust. The most common are an upraised fold, or anticline, and a downfold or syncline; while a fault (basically a vertical landslip) can be classified as a step-fault (self-explanatory), a horst with an upraised bed, or a rift or graben, with a depressed bed.

The landscape is also carved by various forms of erosion or denudation to various extents; igneous rocks being much more resistant than sedimentary. Limestone, for instance, is fairly rapidly dissolved by rain-water, which can thus hollow out huge, weird and beautiful systems of caves, in which stalactites and stalagmites grow from roof and floor respectively, formed from dripping lime water.

Ice and glaciation also do their slow work: water freezes in cracks in rocks and, expanding, breaks the rock apart. Glaciers gouge out U-shaped valleys, smoothing and polishing hard rocks in their inexorable progress, while the polar ice caps themselves constantly affect the Earth's weather patterns.

Streams and rivers erode away their banks, sometimes flood the land and occasionally form spectacular waterfalls – and altogether carry 8 billion tonnes of salts to the sea each day. Each kilogram of sea-water contains 35 grams of dissolved salts of all kinds, the chief one being sodium chloride. The sea itself constantly batters at the coasts and transports vast quantities of material from one place to another.

The Pacific Ocean is twice as large as the Atlantic and covers some 30 per cent of the surface area of Earth, while oceans cover a total of 70 per cent altogether. (Our planet would be more properly named 'Water'!) Over 99.3 per cent of Earth's total water is accounted for by the oceans, ice caps and glaciers, all other forms – lakes, rivers, inland seas, springs, streams, ponds, snow and

■ Planet Earth and its Moon, seen from a small asteroid making a close approach. Only the lights of cities on the night side (especially, here, North America) reveal the presence of Man.

ice on mountains, water vapour in the air, or water simply contained in the ground – making up the remaining 0.6 per cent or so.

The oceans today have an average depth of less than 4,000 metres, but can be as deep as 11,000 (at the Marianas Trench off the Philippines), which would more than submerge Mount Everest. Indeed, if all the water were removed, the mid-ocean ridges would be more spectacular than any of the continental mountain ranges. Under the ocean floor, in relatively shallow shelf regions, lie large quantities of oil and gas, as well as phosphates (used for fertilisers), sand and gravel, and shell sand (for the cement industry).

In the future the oceans are certain to be exploited to a much greater extent, since they contain almost every element in some fraction. Potato-sized nodules of manganese ore lie on the bed of the Atlantic Ocean, though in water too deep for it to be yet economically viable to collect them. Another major industry is based on fish, which are in turn dependent upon the shoals of tiny plankton (forms of which could in future be processed *directly* into food). Energy, too, may one day be produced by utilising the differences in temperature between the solar-heated surface layers and colder layers below.

The ocean tides are caused by the Moon's gravity as it swings around its orbit, and to a lesser extent by the Sun. Waves are caused mainly by winds blowing across open sea. Both tides and waves have energy-producing applications, whereas ocean currents, warm or cold, have a considerable effect on the climates of countries and islands near, which they flow: the Gulf Stream is a notable example. Indeed, the oceans have an enormous effect on the world's weather and climate in general, as air is warmed or cooled by the water it passes over, and on the way picking up moisture in vast amounts.

The atmosphere, far from being a static blanket around our planet, is a dynamic, constantly moving system, and although it is only as thick, relatively, as the skin of an apple, all life depends upon it. Although it is about 160km thick, it is in the bottom 5km of the troposphere that nearly all the clouds and weather is contained. At the tropopause, the boundary between the troposphere and the stratosphere at about 13km, are a number of zig-zagging jet streams which, apart from being sought out by airline pilots travelling in the same direction (the air in them moves at up to 600km/hr), play an important part in forming the fronts and depressions that are such a feature of our weather. The next layer is the

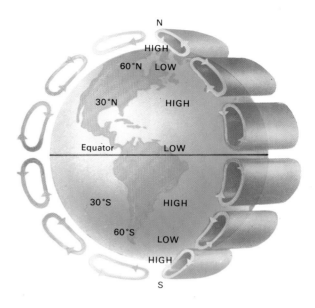

■ ABOVE: Rising hot air and falling cool air cause patterns of circulation known as Hadley Cells, after George Hadley a London lawyer who first proposed the process in 1735.

mesosphere (between 50 and 80km), where the air is very rarefied, and above this is the ionosphere, the exosphere (above 450km) and interplanetary space.

The composition of air is (apart from water vapour): nitrogen 78 per cent, oxygen 21 per cent, argon 0.9 per cent, carbon dioxide 0.03 per cent, plus some hydrogen, helium and other gases. At least 20,000 tonnes of meteoric dust drift into the atmosphere each year, adding to the dust from many other natural and man-made sources.

At sea level the atmospheric pressure is usually around 1,000 millibars (1mb equals 1.02 grams per square centimetre); at 5,000 metres, this figure has halved, and at 16,000m is only one-tenth. When heated air expands and rises it creates a low-pressure system. The air that rushes in to replace it causes winds. On a large scale this process is repeated between hot equator and cold poles, forming several Hadley Cells (see diagram). Conversely, where cooler air sinks a high-pressure area is created; the largest such belt occurs around latitude 30° north and 30° south, and is known as the 'Horse Latitudes'.

The Earth today has an equatorial diameter of 12,756km, a polar diameter of 12,713km. Its equator is inclined to its orbit by 23° 27', which accounts for its seasons (the Sun's rays do not strike each hemisphere with the same intensity). Its average distance from the Sun is 149,600,000km. To any extraterrestrial visitor Earth must seem a highly active planet – beneath its surface, in its oceans and its atmosphere and beyond. But on this smoothly running, self-regulating planet there exists one species which appears to regard the Earth as its personal property, to use or abuse often with no apparent thought for future consequences. That species is of course *Homo sapiens* – modern Man.

34

# Man and
# the Universe

Some astronomical co-ordinates:
- Celestial sphere: imaginary dome of the sky, 180° from horizon to horizon.
- Ecliptic: path of the Sun among the stars/plane of the Earth's orbit.
- Celestial Equator: projection of Earth's equator onto celestial sphere.
- Celestial Poles: projection of Earth's poles onto celestial sphere.

*'Living creatures arose from the moist element as it was evaporated by the sun. Man was like another animal, namely, a fish in the beginning.'*

ANAXIMANDER OF MILETUS (6th Century B.C.)

Seventy million years ago a new era in Earth's history began – the Cenozoic, or new life, era. It was during the early stages of the Cenozoic that tree-living mammals developed which, according to many authorities, became the primates and the direct ancestors of man. By 25 million years ago, in the Miocene period, small gibbon-like proto-apes had developed and by 5 million years ago the earliest man-like primate *Australopithecus,* walked upright on the plains of East and South Africa. He used tools, and is generally considered to be Man's true ancestor – the 'ape-man'.

Originally vegetarian, the Australopithicines stood upright to pick leaves and fruit from trees, and to look around for the possible approach of enemies; they could also use sticks and stones as weapons. As they gradually spent more time walking on their hind limbs their forelimbs and hands were left free to assist the gathering of food and fashioning of implements. Stones chipped in a way that could not possibly be natural have been found in dated rock strata in Tanzania. These are the axes and knives of *Australopithecus habilis,* who lived 2 million years ago, and are the first known *shaped* tools.

The line of toolmakers continued until, between 1 million and 500,000 years ago, *Homo erectus* arrived – some of them in Europe.

They stood about 1.5m tall, and had long, straight leg bones, so could walk as well as run upright, unlike their predecessors. After so long on the plains their feet lost the need for an opposing, gripping thumb like their hands. But the most important development was the brain capacity of *Homo erectus,* which doubled from 500 to 1,000cc. The discovery and use of fire contributed greatly to their survival through the Ice Age that began to grip Europe 600,000 years ago; and it is at least possible that they had simple speech. By 200,000 years ago *Homo sapiens* – which means 'thinking Man' – had begun to

replace other forms of man that had evolved seemingly on parallel lines to each other. *Homo sapiens neanderthalis,* Neanderthal man, clothed himself in skins 100,000 years ago, but by 35,000 years ago he, too, was replaced – by Cro-Magnon man. Cro-Magnon's clothing and tools were sophisticated: furs, shoes made of skin, cleverly made spears and harpoons. He had a spoken language – and he had become an artist, producing cave paintings and carved ornaments. He was the first 'modern man': *Homo sapiens sapiens.*

'... the newly observed deviation from a normal standard of human structure is not in a casual or random direction, but just what might have been anticipated if the laws of variation were just such as the transmutationists require.'
*Antiquity of Man*
Charles Lyell

By 10,000 years ago Mesolithic man fished with nets, hunted with bows and arrows, travelled by dug-out canoe, and lived in communities on the banks of rivers or near the coast. True agriculture, including growing crops from seed and animal husbandry, was developed by Neolithic man 8,000 years ago, so he was no longer dependent on hunting and tended to live a more settled existence in larger, well-ordered communities.

For a very long time early man had no knowledge of the universe apart from his immediate environs. He saw the sky, naturally, and it is possible that he noted the connection between the movements of the heavenly bodies and the seasons on Earth, and so began to use the connection as a rudimentary calendar. He regarded with awe such events as volcanic eruptions, earthquakes and thunderstorms, and began to attribute them to the powers of gods. He had no way of differentiating be-

tween atmospheric phenomena such as clouds, rain, lightning, rainbows, aurorae, meteors, etc. and objects beyond the Earth: the Sun, Moon, eclipses, Milky Way, stars, and comets.

There is evidence that Man has had religions since the time of Cro-Magnon. For thousands of years the only answers to that profound question: "Where did we come from?" were theological and mythological.

As late as the 17th and 18th centuries, flint axe-heads and the bones and teeth of animals which obviously no longer existed were an embarrassment, especially to those whose religious beliefs were based on the Bible. A book published in 1655 by a Frenchman, Issaac de la Peyrère, was publicly burned because it suggested that strangely chipped stones which he had collected had been made by primitive men – *before* the time of Adam. J. F. Esper in Germany; J. Frere and Father J. McEnery in England; P. C. Schmerling in Belgium; J. B. de Perthes in France; all found fossil evidence, but few would read their books or listen to their interpretation of the evidence.

The man who opened the eyes and ears of Victorian England, and eventually the world, was the naturalist Charles Darwin, whose book *On the Origin of Species,* published in 1859, put forward the theory of 'natural selection' as the mechanism by which today's many varieties of plants and animals, occupying so many different niches of existence, came about from common ancestors. Although it did not mention Man, the book was received with shock and horror. Darwin had been impressed by an earlier work by the Scottish geologist, Charles Lyell, called *The Geological Evidence for the Antiquity of Man,* which proposed basically that the forces which carve and change the landscape today must always have been at work – and for much longer than the 6,000 years then considered to have elapsed since Earth's creation. When this was published Darwin was sailing as a naturalist aboard the *Beagle,* a trip which took him to the Galápagos Islands in the Pacific. Each of these relatively new, volcanic islands was a tiny, self-contained world in which evolution had produced obviously related yet differing species.

The English biologist Thomas Huxley's book *Man's Place in Nature* (1863) pre-empted Darwin by comparing Man with the living apes anatomically, but Darwin's *The Descent of Man,* published in 1871, raised further storms of controversy. Most people took offence at what they thought both books were implying: that their own, personal, direct ancestors were hairy apes. It did not make the idea of evolution easy to accept.

'I was looking through Man's Place in Nature the other day. I do not think there is a word I need delete, nor anything I need add except in confirmation and extention of the doctrine there laid down.'
T. H. Huxley

The wealth of fossil evidence uncovered since Darwin's time has vindicated him. His great theory of evolution has meanwhile undergone modification; discussions and discoveries continue, and our ideas on some aspects may yet have to change. It is only quite recently that the theory was advanced that the dinosaurs, rather than being cold-blooded, slow-moving reptiles, were warm-blooded, energetic – and, in fact, the direct ancestors of birds. Yet we still do not know for certain why these creatures died out so suddenly 65 million years ago.

Man has often been more disposed to speculate about the nature of the universe than about his own origins, though. The key to the beginnings of science and astronomy is, of course, *change.* If the planets did not change their position among the stars, if the Moon did not change its phases, no-one would pay them much attention. If counting the cycles of the Moon, or the regular reappearance of the Sun or a star at a certain point on the horizon at the same time each year, enabled Neolithic man to know when to plant a crop, say, there was a practical reason for him to do so.

But Man is first and foremost a curious animal, and merely learning to tell the date and time would not satisfy him intellectually for long; on the other hand, being able to do those things accurately was essential when he came to make more sophisticated observations later.

Because ancient people considered themselves and their world to be the centre of the universe they inevitably came around to believing that changes in the sky must mean changes on Earth were to follow, so astrology was born; and until quite recently there was no sharp dividing line between it and the true science of astronomy. Of particular value to both was the Zodiac, that narrow belt along which the Moon and planets (and Sun, invisibly) travel. The Mesopotamians used this as early as 3,000 B.C. Astrologers, even today, would have us believe that a change in our life will occur when a planet moves from one 'house' of the Zodiac to another – even though the 'sign' under which we were born is no longer where it was when the Greek Claudius Ptolemaeus (Ptolemy) wrote the standard textbook of astrology/astronomy around A.D. 140! In fact, each Zodiacal sign has moved back 24 days since then.

Ptolemy was also responsible for the best-known geocentric, or Earth-centred, model of the Solar System, and indeed the universe. Working to some extent on the excellent observational astronomy of Hipparchus (about 150 B.C.), he explained in his great work *The Almagest* the motions of the planets and other bodies by means of 'deferents' and 'epicycles', as in the diagram, assuming that they must move in perfect circles. He was not the first to propose a system of this type. Eudoxus and Aristotle had both attempted to solve the puzzle but Ptolemy's system was the first to explain the motions within the observational limits of the time.

However, as hundreds of years passed, observations revealed increasing errors in the predicted positions of the planets, and many astronomers became dissatisfied. More small 'epicyclets' were added by some, and the whole business became totally unwieldy.

It was a Polish canon, Nikolaus Copernicus, who had the temerity to suggest that instead of all bodies revolving around the Earth, they (Earth included) went around the Sun. He set out his theory in his book *De Revolutionibus Orbium Coelestium,* published in 1543 in the last weeks of his life; a deliberate move, since he was already ill, and knew quite well that the Church would bitterly oppose such a proposition.

In 1600 a brilliant young mathematician, Johannes Kepler, joined the great observational astronomer Tycho Brahe as assistant. After Tycho's death a year later in 1601 Kepler concentrated his investigations on the retrograde movements of Mars, which were proving so difficult to explain. Using Tycho's records, and by calculation and guesswork, he found that its orbit was not the circle

everyone before had believed – but an *ellipse.* (Even Copernicus had had to resort to small epicycles to account for some discrepancies with observations.)

An ellipse has not one focus, like a circle, but two foci, and Kepler found that the Sun occupied one of these, the other being empty. This fact is now known as 'Kepler's First Law'. His next great discovery was that a line drawn from the Sun to a planet sweeps out equal areas in equal times. This means that a planet moves faster when it is closest to the Sun, or at perihelion, than in the outer part of its orbit or aphelion. That is his Second Law. The Third Law enables the mean distance of any planet from the Sun to be calculated from its revolution period.

Despite these advances, Kepler was a mystic and believed in the 'music of the spheres' – each planet emitted a musical note. He wrote a book which is often quoted as one of the first works of science fiction: *Somnium* ('The Dream'), about a visit to the Moon.

Kepler was able to describe the Moon as a world, with its own mountains and valleys, and that was not fiction but due to the greatest advance in practical astronomy to date – the invention of the telescope. Kepler was in correspondence with Galileo Galilei who, although he did not invent the refracting telescope, was responsible for its first use as an astronomical instrument. He wrote to Kepler in 1597 that he had accepted the Copernican system (though had been afraid to admit it).

'There is no fundamental difference between man and the higher mammals in their mental faculties.'
*Descent of Man*
Charles Darwin

Galileo had heard of the Dutch discovery of the telescope while visiting Venice in 1609, and very quickly made one from spectacle lenses – one convex one concave – which magnified about three times. Later he ground his own lenses, and his best instrument could enlarge 30 times.

He turned it skyward in 1610, and later published his amazing discoveries in a book *Siderius Nuncius* (Sidereal Messenger). This produced reactions of excitement in some quarters, disbelief in others, and displeasure from the Catholic Church, for Galileo's observations clinched the Copernican theory. Venus appeared in his telescope as a crescent and later in other phases, like

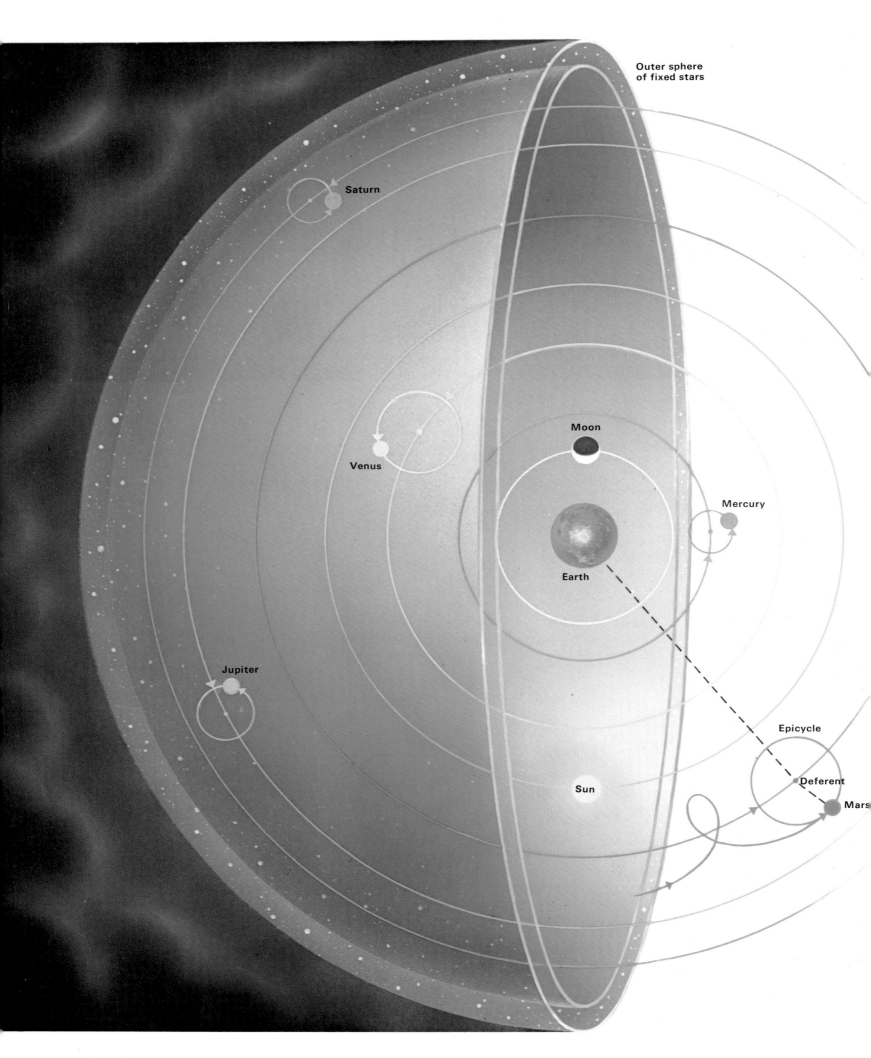

**Outer sphere of fixed stars**

Saturn

Venus

Moon

Mercury

Earth

Jupiter

Sun

Epicycle

Deferent

Mars

■ **The universe according to Ptolemy,** who was a scholar at the Alexandrian Library in Egypt around A.D. 140. His famous 13-volume work on astronomy *The Almagest* (an Arabic form of the Greek description of the work as *megiste* 'the greatest') was originally entitled *The Mathematical Collection.* Ptolemy's system of epicycles and deferents at least allowed the prediction of the positions of the planets and other bodies of the Solar System. At lower right the retrograde movement of Mars is 'explained' by this method. But the positions of Mercury and Venus soon had to be restricted to within 47° of the Sun to account for their observed motions.

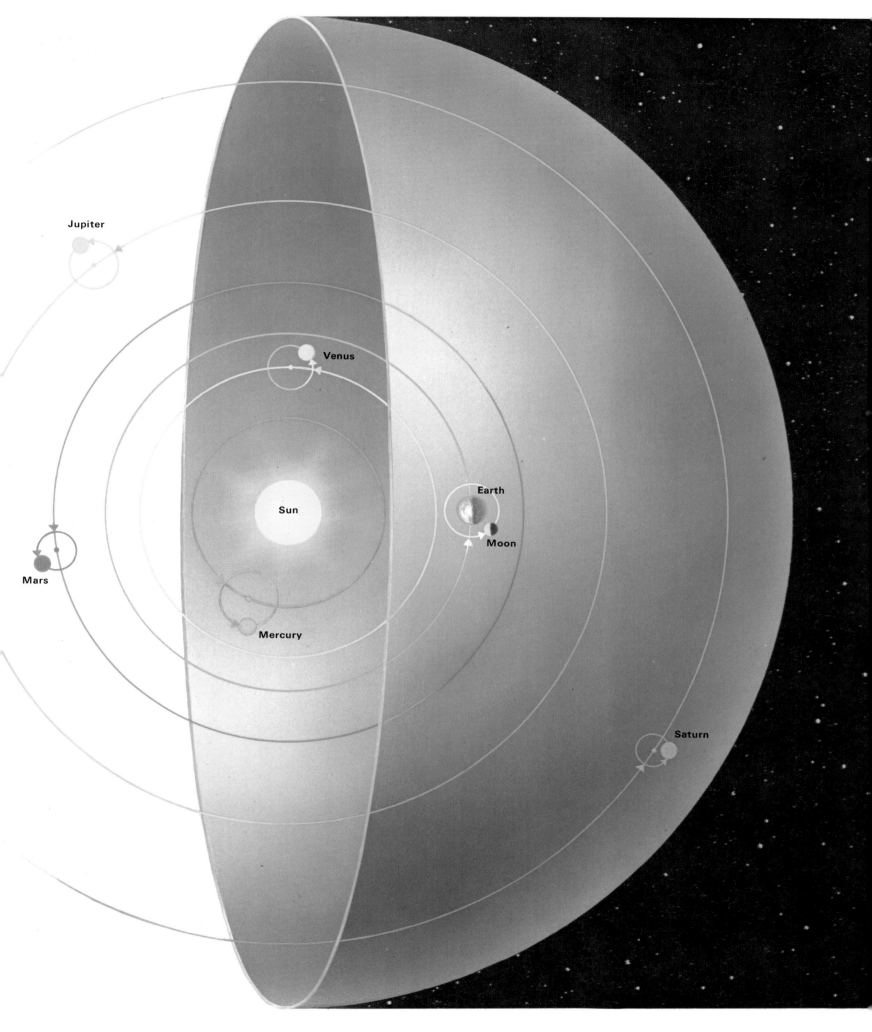

Jupiter

Venus

Sun

Earth

Moon

Mars

Mercury

Saturn

■ **The Copernican system.** The printer of the book *De Revolutionibus* in which this was proposed, a Lutheran minister, added the words *Orbium Coelestium* ('of Celestial Orbs') in an apparent attempt to suggest that the Earth was not included. He also wrote a preface to explain that it was intended chiefly as a useful method of calculating the positions of planetary bodies more accurately than previous systems, such as Ptolemy's, permitted. But even so it was strongly criticised. As Martin Luther himself had pointed out in 1539: 'Joshua bade the *Sun*, not the Earth, to stand still.'

the Moon; Jupiter was seen to be attended by four small satellites whose positions changed from night to night, so that they obviously circled it like a miniature solar system. As for the Moon, it was clearly a world, with its mountains and craters casting deep shadows, and dark patches which Galileo assumed were seas.

Despite Galileo's attempts to convince churchmen of the reality of his discoveries he was condemned by the Inquisition (on the evidence of a false document) and imprisoned (the sentence was commuted by the Pope to house arrest: another Pope, John Paul II, revoked the condemnation in 1980).

Another follower of Copernicus, who also ground lenses and made excellent telescopes, was the Dutch mathematician and astronomer Christian Huygens. He lived in Holland where, unlike Galileo, he was honoured for his discoveries. While Galileo was puzzled by the appearance of Saturn, which seemed to him to have handles, Huygens correctly identified them as flat rings which do not touch the planet. He also discovered Saturn's largest satellite, Titan, and found that Mars has a 'day' very similar in length to ours by observing certain dark markings which rotated with the planet. He also believed that other stars were likely to have planets of their own, which might be inhabited.

Kepler introduced a small convex lens as his eyepiece, which gave an inverted view that enlarged more and gave an image that could be projected. But in 1668 the English scientist Isaac Newton

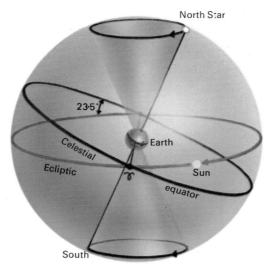

■ Precession of the Earth's axis means that the position of the celestial north pole gradually changes. The celestial equator can be seen to be inclined by about 23.5° to the ecliptic. (The Zodiac extends a few degrees to either side of this.) The nodes are shown, the closer still known as 'First Point of Aries'—though, due to precession, the spring equinox on March 21 now falls in the stars of Pisces!

introduced a new type of telescope altogether. Apart from being responsible for a prodigious output of writings on mathematics, the nature of gravity and even (later) theology, he investigated the nature of optics.

Two years before, he had split sunlight into a spectrum; in trying to construct a telescope whose image was free of rainbow-coloured fringes – chromatic aberration – he concluded (incorrectly) that no lenses could ever yield colour-free results. He then investigated the properties of curved mirrors and made a successful small reflecting telescope, with a mirror one inch (25mm) in diameter. In 1671 he presented a model of his reflector to the

Royal Society, which promptly elected him a Fellow. The 'Newtonian reflector' is still popular today with amateur astronomers, usually with a 150mm main mirror.

Newton's researches on gravity led to the discovery of inertia – the law by which objects moving in a straight line tend to continue to do so unless deflected by some outside influence. It occurred to Newton that the Moon should obey this law and fly away in a straight line, unless some invisible force were constantly pulling it toward Earth, converting its path into a circle (or ellipse). That force was gravity.

The same law applies to a planet or comet moving around the Sun, of course – a comet's orbit being highly elliptical. By employing Kepler's Third Law, Newton was able to show that 'every particle in the universe attracts every other particle with a force proportional to the product of their masses, and inversely proportional to the square of the distance between them'. In other words the pull of gravity between two objects decreases with distance. It will be only a quarter as strong if they are moved twice as far apart, one-sixteenth if four times as far, and so on.

Conversely, it explains Kepler's Second Law – why a body moves more slowly in the more distant part of its orbit: the farther away, the weaker the pull of gravity from the Sun. In the case of a sphere, its attraction on a particle outside itself is as if the entire mass of the sphere were concentrated at its centre, Newton said in his book, the

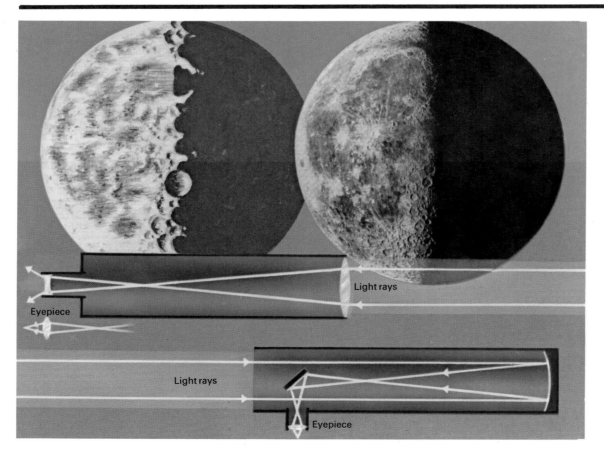

■ TOP: Galileo's first drawing of the Moon, compared with a modern photograph, also taken near third quarter. ■ BELOW LEFT: The principle of Galileo's telescope, showing how the light rays are brought to a focus, then made to diverge by a concave lens in the eyepiece. Alongside is the effect of a convex or converging eyepiece lens as introduced by Kepler. The lower drawing shows the principle of the reflecting Newtonian telescope.

■ RIGHT: NASA's Space Telescope. Designed to be placed in Earth-orbit by the Shuttle, it will allow observations in the entire range of wavelengths, most of which are absorbed by Earth's atmosphere. It will also enable astronomers to obtain images of objects seven times farther away and 50 times fainter than ground-based instruments permit.

*Principia,* published in 1687. He also explained the tides, and showed that the Earth is not exactly spherical but has an equatorial bulge due to the centrifugal effect of its rotation (confirming measurements made at different latitudes). Newton went on to calculate the effects of the planets' mutual perturbations on each other, and compared their predicted positions with those observed. Although he was undoubtedly a genius, Newton, like Kepler, had his odd side. He was an 'alchemist' and searched for the 'elixir' of eternal life. He had many quarrels with the Astronomer Royal, the Reverend John Flamsteed, over the publication of a new and much-needed star catalogue – the first since that of Tycho Brahe.

When Flamsteed died in 1719, his position at Greenwich was filled by Edmond Halley. Halley, who is usually best remembered by the comet that bears his name, had been largely responsible for the financing and publication of Newton's *Principia.* He made many advances in astronomy. He studied the motions of the Moon for nearly 20 years, and found a method to measure the distance of the Sun from the Earth by using transits (their passage across the face of the Sun) of the planets Mercury and Venus, although he never saw one of Venus.

During Halley's time the lenses of the refractor telescope were improved by the English optician John Dollond, who combined crown and flint glass to give achromatic (colour-free) results, despite Newton's pessimism. The reflector did not come into its own until a Hanoverian musician called Friedrich Wilhelm Herschel took up amateur astronomy around 1771. After 200 abortive attempts he produced a main mirror or 'speculum' (at that time the curved mirror was made of an alloy called 'speculum metal', not glass) about 130mm across. He went on to make much larger mirrors and in 1781, while engaged on a 'star count' in selected areas of the sky, discovered a greenish disc that puzzled him.

## A Seventh Planet

At first he took it for a comet; it later proved to be a new planet – Uranus. The discovery caused tremendous excitement; no one had considered that there might be a planet out *beyond Saturn,* the outermost. Sir William Herschel (as he became in 1816) constructed telescopes with mirrors up to 4 feet (122cm) in diameter, about the largest for those with a metal speculum. Once glass with a silver, and later aluminium layer, deposited electrically, came into use, mirrors could be made larger.

The 100-inch (250cm) telescope at Mount Wilson, California, was completed in 1918, and it was to remain the largest in the world for many years, the 200-inch at Mount Palomar, was started in 1928 but, due to World War II, not completed until over 20 years later. It was supplanted as 'world's largest' by the 6-metre, 42-tonne mirror now used in the Caucasus by Soviet astronomers.

An alternative system to Newton's was devised by a French sculptor, N. Cassegrain, in 1672. Instead of being reflected out of the side of the tube by a flat mirror, the light-beam is reflected by a small convex mirror back down the tube and out through a small hole in the centre of the speculum. It is this system that is used at Palomar, and in NASA's new Space Telescope, to be launched by the Shuttle during the 1980s. The Space Telescope has a mirror only 2.4m across, but because it will operate outside our murky atmosphere it should give images more than ten times clearer than ground-based instruments. It would be able to resolve a 10p coin 600km away and resolve objects 300km across on Jupiter – equivalent to the Pioneer fly-by missions.

Orbital telescopes are of great value, not just for observations in visible light but for those in other radiations. Just as waves from infra-red to microwaves are absorbed by carbon dioxide and water molecules, so other wavelengths are blocked by other constituents of our atmosphere. Astronomers need *all* wavelengths in order to build up a complete picture of the universe; it is only by using space-based instruments that they can obtain this.

Obviously, only electronic sensors and other automatic recording instruments can be used in such satellites and probes at present. A whole battery of new detectors have come into use. But most astronomers have a secret ambition to work on an orbiting observatory, or one on the only other world from which observations have yet been made – the Moon.

- Apparent size from Earth: ½° (30 seconds) on celestial sphere.
- Mean diameter: 3,476km. Escape velocity: 2.4km/s.
- Rotation period: 27.3 days.
- Sidereal period: 27.3 days.
- Mean distance from Earth: 384,000km.
- Inclination of orbit (to ecliptic): 5.15°.
- Inclination of equator to orbit: 6.73°.
- Mean density: 3.34.
- Surface temperature: day max. 110°C: night min. −173°C.

*He bade the moon come forth; entrusted night to her; made her a creature of the dark, to measure time; and every month, unfailingly, adorned her with a crown.*

Enuma elish (2nd millennium BC Mesopotamian creation myth)

There is no general agreement on how the Moon itself was formed. In the 19th century the 'tidal theory' proposed by George (the astronomer son of Charles) Darwin was popular, and remained so for many years. According to this the rapidly rotating early Earth became pear-shaped, and the narrow end broke away to become the Moon. There are serious mathematical objections to this, but supporters of 'modernised' versions suggest that the heavy iron accumulating in the Earth's core caused it to spin more rapidly, to one rotation in only just over 2½ hours, at which point the equatorial bulge deformed into an egg shape. Eventually the rapid spinning motion caused the bulge to break away. However, Moon rocks returned to Earth do not seem to support the theory, as their chemical composition differs from the rocks in Earth's crust.

Another theory, which at least explains the different composition, claims that the Moon formed elsewhere and was captured by Earth's gravity; or it may have consisted of several 'moonlets' which were captured and then aggregated into one large body. The Moon is in fact unusually large as a satellite. Although not as large as some of the moons of Jupiter, Saturn or Neptune, it is only about 80 times less massive than Earth, while the other satellites have less than a thousandth of the mass of their parents. For this reason the Earth-Moon system is often referred to as a 'double planet'.

The theory that the Moon was formed of particles in the solar nebula near the Earth – a small-scale version of the process that formed the planets around the Sun – is more widely accepted. The different chemical composition of lunar rocks still poses problems, especially their low density and the lack of the iron that Earth possesses. Possibly a combination of these theories is the answer. A small body may have been captured by the early, still-differentiating Earth, then collected around it the left-over debris of mantle-like material in orbit and grown larger.

Or a large piece of debris may have ploughed into the Earth at an oblique angle, spraying low-density crustal material, or even some mantle, into space. This then formed a ring of hot dust around the Earth; its volatile substances evaporated, leaving silicates with some metals such as aluminium and titanium (which are found in lunar rocks) and these condensed into the Moon. One item on which there is agreement is that the Moon formed about 4.6 billion years ago – that is, at the same time as the rest of the Solar System – and was originally much closer to the Earth than it is now (see illustration on page 27), but spiralled outward fairly quickly. Earth's gravitational influence on the Moon then slowed it down until one face turned permanently toward Earth in synchronous rotation.

---

'Oh! Swear not by the moon,
The inconstant moon,
That monthly changes in her circled orb ...'
*Romeo and Juliet*

---

The conservation of angular momentum in the Solar System applies equally to the Earth-Moon system: the mutual effects of gravity and rotation are slowing it down, as the energy of tidal friction is dissipated as heat. The result is that they are very slowly moving apart, at about 3cm per year. One day, billions of years hence, when they are about 500,000km from each other, the rotation period of both Earth and Moon, and its revolution period, will all be 45 days and *both* bodies will keep the same face always turned to each other.

The tides caused by the Moon on our oceans are obviously much easier to detect than Earth's gravitational effect on the rocks of the Moon, but the Moon *does* have a slight permanent bulge on the side facing Earth. The Moon's effect on the ocean tides means that most parts of the world experience maximum, or spring tides twice every lunar month, at New and Full Moon, when the Sun, Earth and Moon are in line. In the first and last quarter of the Moon, the Sun, Moon and Earth are at right angles, thus the pull of the Sun counteracts the Moon's gravity so that tides are smaller, known as neap tides.

The phenomenon of an eclipse takes place when either the moon lies between the Sun and Earth (solar eclipse) or the Earth lies between the Sun and Moon (lunar eclipse). But why, then, is there not an eclipse both of the Sun and Moon every month? The reason is that the Moon's orbit is tilted by an angle of 5° to Earth's (i.e. the plane of the ecliptic). It is only when the Moon is near to, or crossing the plane of the ecliptic, at a point called the nodes – one at each side of its orbit – and is at Full Moon or New Moon, that a lunar or solar eclipse respectively occurs.

Since New Moon occurs every 29.5 days and the nodes are in line with the Sun only every 346.6 days due to the fact that the Moon's orbit swings slowly round – the precession of the nodes – there is a point at which these cycles coincide, when a sequence of eclipses can be predicted. This is called the Saros cycle, and is 18 years, 11 days and 8 hours in length.

Although the cycle and the prediction of eclipses has been known for some thousands of years, their cause was not known, and they caused great alarm and awe among ancient peoples. All sorts of fantastic reasons were invented to explain why the Moon turned blood red, or into a coppery shield, or even vanished altogether.

If the Moon is at the far part of its orbit (apogee – the closest point is perigee) when a solar eclipse takes place, the result is an annular eclipse. The Moon's disc appears too small to cover the Sun completely, and a ring of sunlight remains visible around the

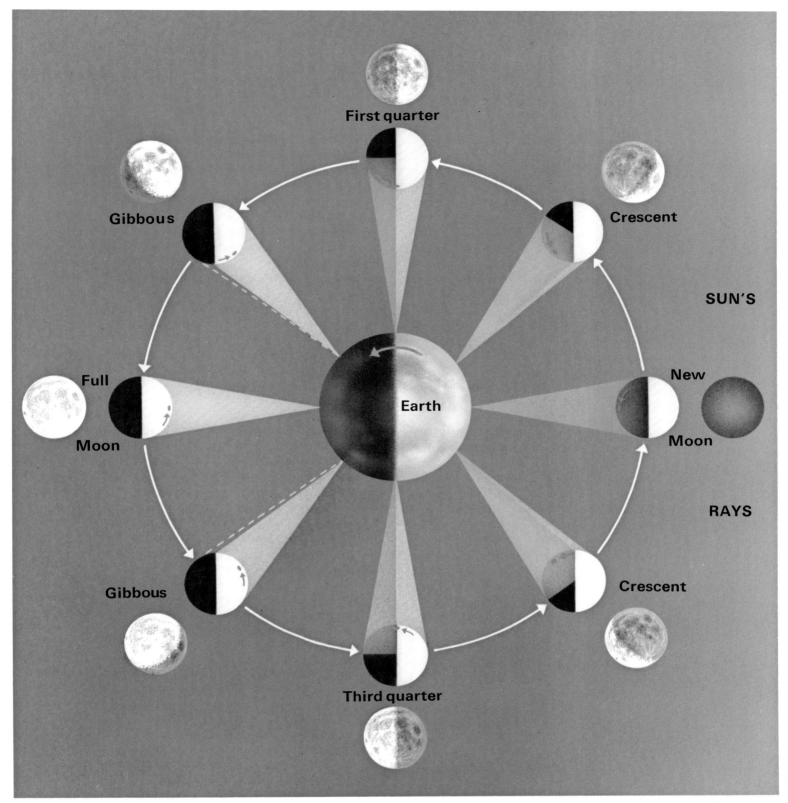

First quarter

Gibbous

Crescent

SUN'S

Full

New

Moon

Moon

RAYS

Gibbous

Crescent

Third quarter

■ ABOVE: The Moon's phases. The outer ring shows the illuminated Moon as it appears to an observer on Earth. Earthlight reflected onto the Moon is also shown; the phase of the Earth as seen from the Moon is always exactly opposite to that of the Moon from Earth (so a crescent Earth would actually shine faintly on the dark portion of the gibbous Moon). The 'sidereal month' of 27.3 days is the Moon's revolution period with respect to a fixed star. The period from New Moon to New Moon (29.5 days) with respect to the Sun, as seen from Earth, is the 'synodic month'.

■ BELOW, LEFT: Orbiter satellites were able to photograph areas of the Moon which are highly foreshortened as seen from Earth. This is an Orbiter 4 picture showing craters near the lunar north pole, from Birmingham (left) to Goldschmidt. BELOW, RIGHT: Partial eclipse of the Moon, as observed by the author, 4.6.74, 21.30 hrs UT (GMT) × 8. The red colour is caused by sunlight refracted or 'bent' round the edge of Earth by its atmosphere.

■ OVERLEAF: The Moon's nearside (p. 42) and farside (p. 43), with landing sites.

Endymion
Atlas
Messala
Hercules
Geminus
MARE FRIGORIS
Aristoteles
Plato
Alpine valley
Eudoxus
LACUS SOMNIORUM
Cleomedes
SINUS RORIS
Posidenius
MARE CRISIUM
SINUS IRIDIUM
Cassini
Caucasus
Le Monnier
Macrobius
L15
MARE
Aristillus
Autolycus
MARE
A17
Archimedes
L2
Linne
SERENITATIS
L17 IMBRIUM
A15
Timocharis
Menelaus
Plinius
Lambert
Haemus
R6
MARE
Tarantius
L20
TRANQUILLITATIS
L16
Prinz
Apennines
Manilius
MARE
Julius
Aristarchus
VAPORUM
Caesar
R8
Langrenus
Herodotus
Eratosthenes
S5
L13
Stadius
SINUS
A11
Vendelinus
OCEANUS
Copernicus
S2
MEDII
SINUS
S6
Theophilus
MARE
Kepler
AESTUUM
S4
A16
Cyrillus
NECTARIS
L8
L7
Hipparchus
Petavius
Encke
L9
A14
Albufeda
Catharina
Fracastorius
L5
Ptolemæus
A12
Albategnius
Furnerius
PROCELLARUM
S3
R9
Alphonsus
Piccolomini
Riccioli
S1
O5
R7
Arzachel
Grimaldi
Thebit
Gassendi
Bullialdus
MARE
Purbach
Regiomontanus
Metius
NUBIUM
Walter
MARE
Pitatus
Deslandres
Fabricius
HUMORUM
S7
Tycho
Maginus
Longomontanus
Schickard
Clavius

Moon, swamping our view of the corona.

New Moon, incidentally, can strictly speaking only be seen during a solar eclipse, as a black disc. Many people refer to the slender crescent visible just after sunset as 'New Moon', but it is then a few days old. At this time the phenomenon known as 'the old Moon in the New Moon's arms' can often be seen; it is caused by 'Earthlight' illuminating the Moon's dark side. An observer on the Moon at that time would see almost 'Full Earth' – the diagram on page 41 shows why. As we know, it was Galileo who thought the dark areas of the Moon were seas – a counterpart of the Earth. Their names,

as first given by the Italian astronomer G. B. Riccioli in 1651, are still in force. The dark plains still retain their old name of seas (*maria,* singular *mare*) with bays (*sinum*); the light regions are continents (*terra*), with marshes (*palus*) and lakes (*lacus*).

## Mapping the Moon

Galileo published his own first observations in *Siderius Nuncius,* together with a crude map. He also measured the heights of lunar mountains by supposed 'triangulation', from their shadows. The first lunar maps of real worth (apart from those of a Spanish cartographer, Langrenus, who used a different system

of names) are those of the German astronomer Johann Hevelius, published in his *Selenographia* in 1647. Riccioli was a colleague of Hevelius, and his own two-volume book used observations by Francisco Grimaldi, who worked with him at the University of Bologna.

Apart from using the Latin names above for the seas and other features, Riccioli named all the most prominent craters after great historical characters such as Plato, Aristarchus, Copernicus, Tycho and many others – as well as his own friends.

Apart from maps by G. D. Cassini (also at Bologna) in 1692 and T. Mayer in 1776, the next major contributions

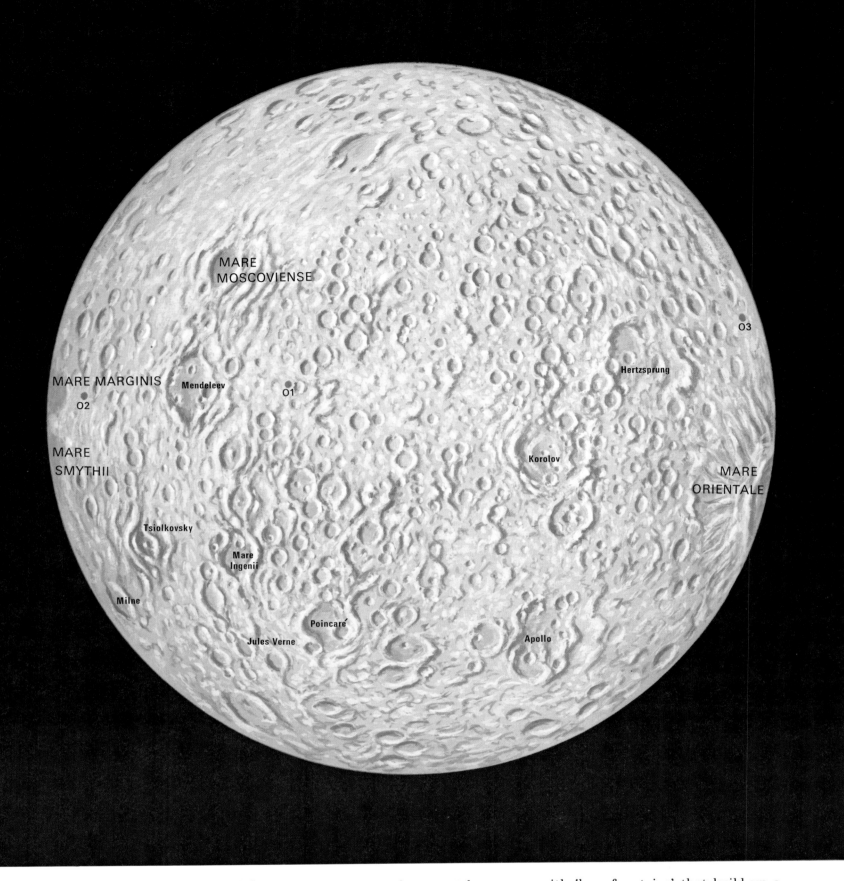

MARE
MOSCOVIENSE

O3

Hertzsprung

MARE MARGINIS       Mendeleev            O1

O2

Korolov

MARE
SMYTHII

MARE
ORIENTALE

Tsiolkovsky

Mare
Ingenii

Milne

Poincaré

Jules Verne                          Apollo

were made by J. H. Schröter, chief magistrate of the town of Lilienthal, Germany. From the observatory he built there he made hundreds of detailed drawings of different features, including his own discovery, 'rilles' or clefts, with a 48cm reflector.

The next good lunar chart was prepared by W. Beer and J. von Mädler (both have craters named after them, though not together) in Berlin in 1837, in a book entitled *Der Mond*. Beer and Mädler, like Schröter, believed the Moon to have an atmosphere. Incidentally, whereas the first lunar maps were drawn with 'north' at the top, like maps of the Earth, later maps put 'south' at

the top. This was because telescopes after Kepler's time reversed the image, and this became the norm for maps. With the advent of Apollo, charts were again turned the right way up.

The Beer and Mädler book was considered the 'standard textbook' on the Moon for many years, and rightly so. My personal favourite in this field is *The Moon* by J. Nasmyth and J. Carpenter, first published in February 1874. It is illustrated with fine drawings and beautiful, correctly lighted models of lunar features, including views as seen by an observer on the surface. It also advocates a rather unique and now outdated theory of crater formation,

with 'lava fountains' that build up a ringwall. By the late 19th century photography began to replace drawings for general accuracy, though not of fine details.

The study of the Moon's topography and mapping is properly called selenography. An international system of nomenclature was not accepted until as late as 1935. The International Astronomical Union (IAU) appointed two scientists, M. A. Blagg from England and K. Müller from Germany, to perform the task of compiling this list, which contained 672 names.

Until that date – and indeed until the 1960s – all the names were on the side of

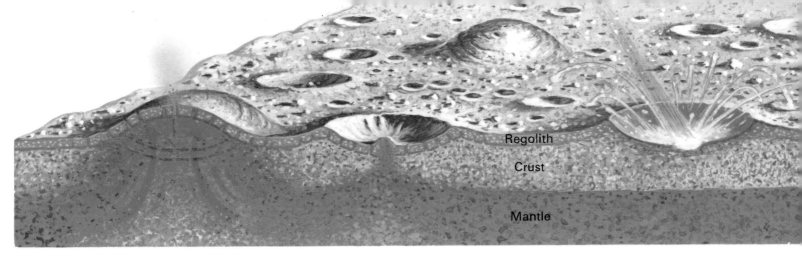

Regolith

Crust

Mantle

the Moon that faces Earth. That does not mean, surprisingly, that only exactly half of the Moon had been mapped. The Moon appears to 'rock' slightly, a movement known as libration, which means that we can actually observe 59 per cent of the total lunar surface from Earth at one time or another, of which 41 per cent is always visible.

### Era of the Rockets

So until little over 20 years ago Man had seen only just over half of the Moon and what he had seen was through telescopes at the bottom of our atmosphere. Then on October 4, 1959, Luna (or Lunik) 3 was launched from Baikonur Cosmodrome in Russia, and passed 7,886km behind the Moon. It took actual photographs of the far side, which it processed on board and then scanned with a TV device. The rather blurred result at least showed that 'farside' was different from the familiar face; hardly any dark, mare regions. Two small dark patches that showed up against the bright highlands were promptly named Mare Moscoviense (Moscow Sea) and Tsiolkovskii (after the Russian pioneer of rocketry), a very dark-floored crater.

Between 1964 and 1968 many Russian and American probes – hard-landers, soft-landers and orbiters – improved considerably on these first pictures, and provided a virtually complete coverage of the lunar surface, except for some polar regions. Almost all of our modern knowledge of the composition of the Moon and its interior has come from these and the manned Apollo missions, which also provided a great many valuable and spectacular photographs.

An examination of the Moon's surface is an especially rewarding experience for the amateur astronomer. If possible it should be viewed through a telescope or even good binoculars, and special attention should be paid to the terminator, the dividing line between the light and dark sides, where the surface is thrown into relief by strong shadows. The first impression given is of thousands of shadow-filled craters. The observer can be forgiven for thinking that these are yawning pits into which a spacecraft could vanish without trace, but this is illusory.

Lunar craters are actually very shallow features with the ground sloping gently up to the rim, then falling only slightly more steeply into a saucer-shaped depression. Anyone standing at the centre of one of the large craters would think that they were on a flat plain as the rim would be below the horizon!

Although the impact theory is widely accepted today (and in view of the widespread cratering of other planets and satellites there seems little doubt of its validity), the craters were originally assumed to be volcanic. Proponents of the volcanic theory, many of whom still put forward modified versions today, made a good case on the basis that craters often occur in 'chains', which are still easier to explain as arising along 'lines of weakness' in the lunar crust; they also point out that craters are not randomly distributed, and that small craters almost always break into large and not *vice versa*. This again is difficult to explain by the impact theory. Patrick Moore has rightly stated that *both* processes must have operated. Lunar rock is in fact basically volcanic in nature, but undeniably meteoric forces have shaped the surface.

Terrestrial volcanoes are in any event usually in the form of a truncated cone – a mountain with a depression in the top. But the 'volcanists' suggest that domes formed, then collapsed, in the plastic surface of the early Moon.

However they were formed, there are tens of thousands of craters over 1km in diameter visible from Earth alone. Of these more than 1,000 are over 16km across, and they continue up in size to giants 240km across and 5km deep – and down to 'microcraters' only millimetres across. Yet it has been estimated that one would now have to wait 100,000 years to see a new crater 1km across appear on the Moon.

Whether the maria have the same origin as the craters is in doubt, although many US astronomers, particularly, seem certain of it. Impact craters were made by projectiles from 1km in size upward, but it would take a body 100km in diameter to gouge out a crater the size of the Mare Imbrium, over 1,000km across. One cannot imagine this happening without disastrous results elsewhere on the Moon – but that does not rule it out as a possibility. The basically circular nature of the maria cannot be denied, even though some have been distorted by later events.

The maria, or plains, consist of mul-

# The Lunar Probes

Coloured dots indicate landing sites on pages 42 and 43.

### 1959

● **Luna 2** : First probe (USSR) to rea Moon. Arrived 13 September, crash between craters Archimedes, Aristil and Autolycus.
**Luna 3** : On 4 October sent back f pictures of farside.

### 1964

● **Ranger 6** : US probe. Reached M on 2 February but TV cameras did work. Crash-landed in Mare Tranq litatis.
● **Ranger 7** : Transmitted 4,308 pho graphs on 31 July; first close-up details down to under 1 metre acro Crashed in Mare Nubium.

### 1965

● **Ranger 8** : Over 7,000 photos s back on 20 February. Crashed in M Tranquillitatis.
● **Ranger 9** : Nearly 6,000 photos details in Alphonsus area, down 25cm, on 24 March. Last of seri
● **Luna 5** : First attempt at soft-landi failed and crashed near Mare Nubi on 12 May.

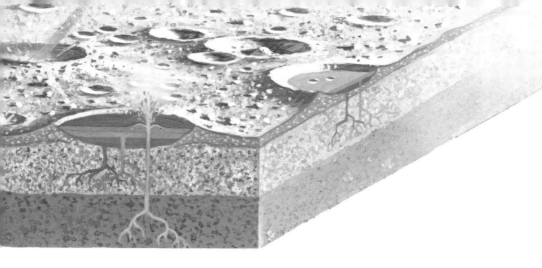

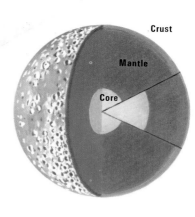

tiple lava flows solidified into basalt, 3–4 billion years old. The cratering process alone would have resulted in the entire Moon looking like farside. But 100 million years after its formation the outer 200km became very hot, and a differentiation process started, with heavier materials sinking and less dense rock rising to the surface of what was virtually an ocean of white-hot, molten rock. This rock, light in colour as well as weight, is now known as anorthosite, and forms most of the solid crust, which is thicker on farside.

## Lunar Rocks

Partial re-melting of this, caused by heat from an unknown source, created another type of rock, KREEP norite, so named because it is rich in potassium (K), rare earth elements (REE) – a group of metals similar to aluminium chemically – and phosphorus (P); or relatively so compared with other rocks. It is abundant in the mountainous regions around the Oceanus Procellarum and Mare Imbrium, and is also rich in thorium and uranium.

Seismic experiments set up by the Apollo astronauts indicate that the anorthositic crust is about 60km deep. The older parts of the surface consist of this rock, converted from smooth sheets by intense cratering into rough, chaotic terrain with an upper layer of powdery 'soil', together termed 'regolith'. In the highlands, where craters cover every available metre, what may once have been imposing, jagged mountains have been overlaid and rounded by 'ejecta blankets' composed of debris thrown up by impacts, and the crust may have been pulverised to a depth of 2km.

By 4 billion years ago the worst of the impact cratering was over, and the large lunar basins – now grey plains, like the Mare Imbrium – seem to have formed at that time. By 3 billion years ago impacts had reached about the present level. This may explain why few rocks older than 4 billion years were found by the astronauts: before the high cratering rate declined all rocks were quickly smashed into dust. If the large basins were formed by late impacts of asteroid-sized bodies they may well have blasted debris – boulder-sized blocks, rubble and dust – for hundreds of kilometres into space, much of it to create secondary craters on its return.

In the final fling, red-hot lava from inside the Moon worked its way to the surface and flooded the floors of the great basins. Whether this was created by slow heating from radioactive isotopes, or was the last remnant of earlier heating is not known, but flow after flow spewed forth and swept across the maria – not in a brief burst, but for up to a billion years. It lapped at the foot-hills of the mountains around Mare Imbrium and Oceanus Procellarum, where its 'frozen' overlapping waves can be seen today on the smooth-looking dark grey basalt plain. The smoothness is deceptive; the plains have been pock-marked by many small impacts since they cooled and solidified, but few major impacts have punctured them.

What of other features? There are domes, which may well be volcanic in origin. There are the bright rays of fine material sprayed forth by some craters, notably Tycho. And there are Schröter's rilles (clefts), which may be mistaken for dried-up river beds, but are actually lava channels. There is no water – or life.

---

7: Unsuccessful soft-landing, October, in Oceanus Procellarum.
8: As Luna 7; 6 December.

### 1966

9: First successful soft-landing, ebruary. Returned four panoramic os of surface in Oceanus Procel- n – resolution down to 1mm.
10: Orbiter – arrived in orbit on ril.
eyor 1: Successful US controlled landing. Transmitted over 11,000 os (some in colour) – details to m – in Oceanus Procellarum on ne.
ter 1: Entered lunar orbit on 14 st. Returned pictures covering 5 million km of surface, then erately crashed on farside in ctober.
11: Inserted in wide elliptical on 27 August.
eyor 2: Failed to make soft- ng on 23 September and crashed Copernicus.
12: Entered highly elliptical orbit on 25 October and returned t (1,100 line) TV pictures.
ter 2: Arrived on 10 November. red nearly 4 million km of surface, ding intended Apollo landing sites. Oblique 'picture of the century' of Copernicus.
● Luna 13: Landed on 24 December in Oceanus Procellarum and sent back three panoramic photos; used a mechanical soil-sampler.

### 1967

● Orbiter 3: Entered orbit on 8 February. Orbit changed several times before deliberately crashed.
● Surveyor 3: Soft-landed in Oceanus Procellarum on 20 April and transmitted 6,320 photos to Earth. Scooped up and photographed lunar 'topsoil'. Very close to Apollo 12 site.
Orbiter 4: Arriving on 8 May, it photographed lunar south pole for first time.
Explorer 35: US orbiting probe to study solar wind, interplanetary magnetic field, radiation etc. in lunar environment. Inserted in lunar orbit on 19 June.
● Surveyor 4: Radio contact lost on 17 July; destroyed by impact.
Orbiter 5: Last of series. Entered orbit on 5 August and photographed not only the Moon but the Earth. Observed a lunar eclipse on 18 October.
● Surveyor 5: Landed safely on 11 September after a serious helium leak was discovered and corrected from Earth. Transmitted over 19,000 photos and made chemical analysis of soil.
● Surveyor 6: Touched down on 10 November in Sinus Medii and took total of 30,000 photos, later ones stereo pairs. On 17 November was first vehicle to lift-off from lunar surface and land again – 2.44m away – to test surface.

### 1968

● Surveyor 7: 10 January. First soft-landing in a highland area, just north of Tycho. Mechanical scoop dug seven trenches, up to 40cm deep. Took 21,000 photos. Last of series.
Luna 14: Orbiter; 10 April. Details never given.
Apollo 8: Orbited Moon at Christmas (21-27 December) and returned to Earth without landing. Astronauts: Borman, Lovell, Anders.

### 1969

Apollo 10: 18-26 May; as Apollo 8 but Lunar Module (LM) made close approach to Apollo 11 site, without landing. Astronauts: Stafford, Cernan, Young.
● Apollo 11: First manned lunar landing on 20 July, at Mare Tranquillitatis. Astronauts: Armstrong, Aldrin, Collins (Command Module).
● Luna 15: Arrived in lunar orbit while Apollo 11 mission was under way. Probably intended to return samples to Earth automatically, but crashed in Mare Crisium.
● Apollo 12: Landed very precisely on 19 November, close to Surveyor 3; removed parts of probe. Astronauts: Conrad, Bean, Gordon (CM).

### 1970

● Luna 16: First automatic collection and return of rock samples. Landed 21 September.
● Luna 17: Landed 17 November; transported Lunokhod 1 – first robotic mobile laboratory, radio-controlled from USSR. Drove 105km in 10 months.

### 1971

● Apollo 14: Landed 5 February. Manual two-wheeled truck used; 43.5kg of samples obtained. Astronauts: Shepard, Mitchell, Roosa (CM).
● Apollo 15: 30 July. Electric-propelled Lunar Rover used for first time. TV camera operated from Earth relayed lift-off from 'Hadley Base'. Astronauts Scott, Irwin, Worden (CM).
Luna 18: Went into orbit on 7 September but failed to soft-land and crashed on 11 September.
Luna 19: Entered close-orbit on 3 October and carried out gravitational measurements etc.

### 1972

● Luna 20: Achieved a landing on 24 February between Mare Crisium and Mare Fecunditatis (mission intended for Luna 18). Drilled for core sample and returned it to Earth.
● Apollo 16: Geological research carried out in highland area north of Descartes by Astronauts Young, Duke, Mattingley (CM). Landed 21 April.
Luna 22: Went into lunar orbit on 2 June.
● Apollo 17: Last Apollo mission. Touched down on 11 December. Highland Taurus-Littrow area. Astronauts: Cernan, Schmitt (first scientist on Moon), Evans (CM).

# Mercury

- Mean diameter: 4,880km.
- Escape velocity: 4.3km/s.
- Rotation period: 58.7 days.
- Sidereal period: 88 days.
- Mean distance from Sun: 58,000,000km.
- Inclination of orbit: 7°.
- Inclination of equator to orbit: 0°.
- Mean density: 5.4.
- Surface temperature: day max. 427°C; night min. −173°C.

*6 Closest planet to the stellar furnace which is the centre of our solar system, one half of little Mercury is for ever bathed in tremendous heat, while the other half, facing away into space, is a paradoxical extreme of bitter cold and perpetual darkness.9*

Worlds in Space MARTIN CAIDIN (1954)

Visually, Mercury is more like the Moon than like any other planet. Indeed, to anyone but an expert many of the images received from Mariner 10 in 1974 cannot be distinguished from photographs of the Moon.

Before 1974, though, our knowledge of the physical features of the planet was slim indeed. In 1889 an Italian astronomer, Giovanni Schiaparelli, compiled the first real map, followed in 1934 by one from the great French planetary observer E. M. Antoniadi. These maps remained standard references until the Mariner 10 probe.

It was generally agreed by the end of the 19th century that Mercury completed one rotation every 88 Earth days, which was also the time it took to orbit the Sun. This meant that it kept one face turned to the Sun. Then, in 1965, two American radio astronomers, R. B. Dyce and G. H. Pettengill, used the giant radio telescope built into a valley at Arecibo to bounce radar waves off Mercury. From their results they were able to establish that it rotates with a period of *59* days, not 88.

The evidence that Mercury does not have an eternally sunward face had been available, had anyone heeded it, since 1962, when radio emissions from Mercury's night side were picked up by astronomers at the University of Michigan. If the dark side were as cold as everyone believed at that time – near absolute zero – there should have been *no* radio waves!

So we have learned an enormous amount since 29 March 1974, when Mariner 10 sped past within 760km of Mercury. It went on to make a complete orbit of the Sun and returned on 21 September 1974, when it was able to photograph new areas.

Mercury's orbit is the most highly elliptical in the Solar System. At perihelion the planet is 46 million kilometres from the Sun, but at aphelion nearly 70 million – an average distance of 58 million km. This means that the size of the Sun varies considerably in Mercury's black sky (black, because there is no atmosphere). The Sun's disc would appear to swell slowly from 2 to 2½ times the size it appears from Earth. The Sun's light and heat would blast down with incredible violence, and of course the whole battery of other radiations would also bombard the unprotected surface. The temperature rises to 600K (427°C), while on the night side Mariner 10 measured temperatures down to 100K (−173°C).

There are two main types of terrain: extremely cratered regions resembling the lunar highlands, and relatively smooth areas similar to the maria. There are also some areas that are much smoother and less cratered than any on the Moon. The craters that are visible range from 100m across to the largest basin, 1,300km in diameter. Like lunar craters they are of all ages, from 'new' ones with sharp rims to barely visible, degraded rings, and some have central peaks. The large craters have ejecta blankets and their walls are terraced, though the ejecta blankets are less extensive than on the Moon – presumably because of Mercury's higher gravity. It has a diameter of 4,880km.

---

'It is the least of all the planets, and lowest, except the moon.'
*A New General English Dictionary*, 1744

---

There are also features not found on the Moon. One such is the Discovery Scarp, one of several ridges that appear to be compression thrust faults. These may have been formed as the core of the planet contracted, perhaps during the process of differentiation. Before the oldest craters were formed, the interior of Mercury had separated out into a very large iron core. 80 per cent of Mercury's mass is contained in this core; it is surrounded by a silicate mantle. Later, there may have been planet-wide volcanic activity, with lava flows that wiped out the chaotic evidence of the accretion process, resulting in the smooth plains between later impact craters. These uncratered areas also reveal shrinkage faults – 'lobate scarps' – not found on the Moon.

As a result of its massive iron core, Mercury has a magnetic field, suggesting that some part at least of the core may be fluid. The field is weak – only a hundredth of Earth's – but even so most scientists had thought that such a field was precluded by Mercury's small size and slow rotation.

Many now believe that the magnetism detected by Mariner 10 is a 'fossil' magnetic field, produced much earlier in the planet's evolution, perhaps when the crust was still capable of movement, and the scarps formed. The result, anyway, is that Mercury has a magnetosphere which deflects the solar wind and forms a cool 'plasma sheet' around the poles, with much more energetic plasma around the magnetic equator.

An outstanding feature on Mercury is the Caloris (from the Latin for heat) Basin, so called because at perihelion, every two years, it is directly beneath the Sun. It was the largest feature seen by Mariner 10, measuring some 1,300km across. The mountains that form its ringwall rise to 2km above the surrounding countryside. Caloris is Mercury's nearest equivalent to the Moon's Mare Imbrium and was probably formed in the same manner. Its floor is unusual for being covered with a 'crazy paving' pattern of ridges and clefts, due to stresses of some sort. But on the side of the planet exactly opposite to Caloris, the terrain is wrinkled into peculiar, corrugated hills. This rippled effect is believed to be due to the focusing of powerful seismic waves through the planet, caused by the Caloris impact.

The Mariner results made it possible to produce proper maps of one hemisphere, at least, for the first time.

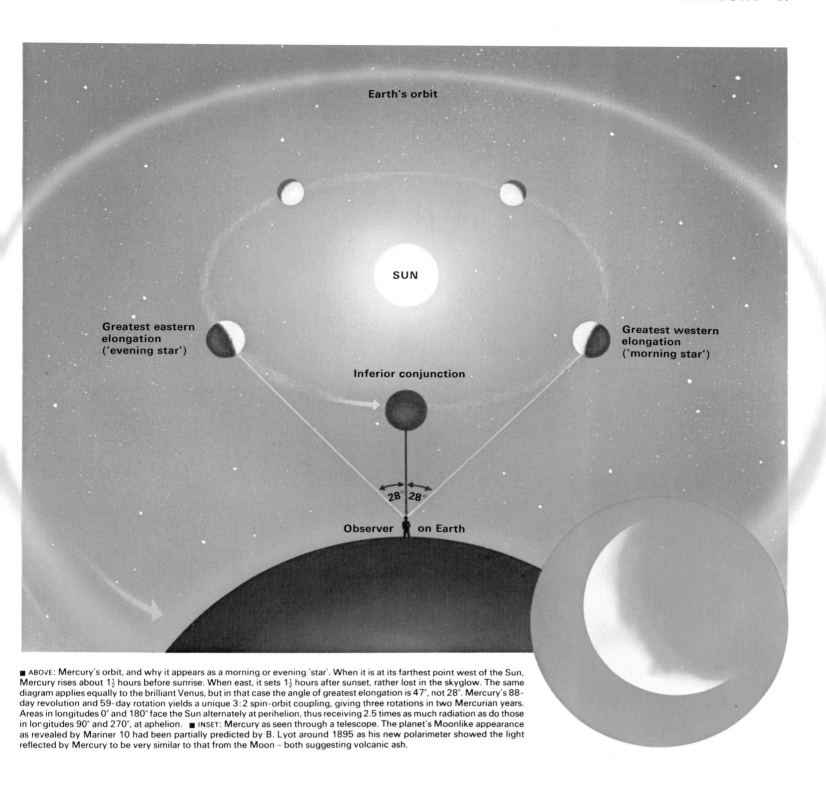

Earth's orbit

SUN

Greatest eastern
elongation
('evening star')

Greatest western
elongation
('morning star')

Inferior conjunction

28° 28°

Observer on Earth

■ ABOVE: Mercury's orbit, and why it appears as a morning or evening 'star'. When it is at its farthest point west of the Sun, Mercury rises about 1½ hours before sunrise. When east, it sets 1½ hours after sunset, rather lost in the skyglow. The same diagram applies equally to the brilliant Venus, but in that case the angle of greatest elongation is 47°, not 28°. Mercury's 88-day revolution and 59-day rotation yields a unique 3:2 spin-orbit coupling, giving three rotations in two Mercurian years. Areas in longitudes 0° and 180° face the Sun alternately at perihelion, thus receiving 2.5 times as much radiation as do those in longitudes 90° and 270°, at aphelion. ■ INSET: Mercury as seen through a telescope. The planet's Moonlike appearance as revealed by Mariner 10 had been partially predicted by B. Lyot around 1895 as his new polarimeter showed the light reflected by Mercury to be very similar to that from the Moon – both suggesting volcanic ash.

■ BELOW: Mercury's magnetic field and magnetosphere. The field is much less strong than Earth's, and the north and south magnetic poles are aligned with Mercury's axis of rotation. As in the case of Earth's, there is a bow shock (1) created by the solar wind, and a magnetopause (2) inside which the lines of force are contained. The Moon has no such field, and the solar wind strikes the surface directly.

■ BELOW: Cross-section of Mercury. The segment of Earth reduced to the same scale shows Mercury's iron core to be much larger, proportionately. It is in fact slightly larger than the Moon, being estimated at 3,600km in diameter.

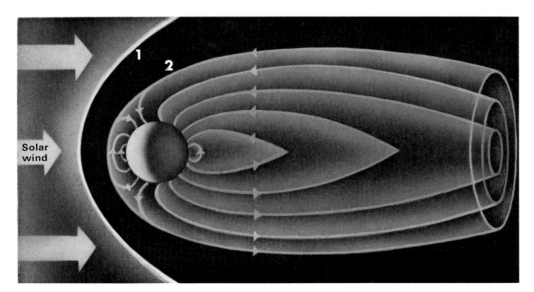

Solar
wind

1
2

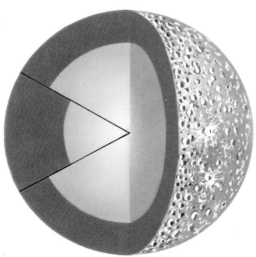

■ LEFT: Two photomosaics of Mercury. The 18 pictures making up the one at the far left were taken when Mariner 10 was 200,000km and six hours away on its approach on 29 March 1974. Two-thirds of this view is of the southern hemisphere. The other mosaic contains 18 pictures taken six hours after Mariner sped past Mercury at 210,000km on the same date. The north pole is at the top and the equator about two-thirds down. Several rayed craters are visible in both views.

■ RIGHT: The huge Caloris Basin, centred at 30°N, 190°W. Note the peculiar intersecting cracks and wrinkles.

■ FAR RIGHT: These Mercurian craters have a distinctly lunar appearance. A 2km-high scarp is visible near the centre of the picture – part of a larger system of faults extending for hundreds of kilometres. In March 1975 Mariner passed as close as 327km. It continues to swing past Mercury every two of its years, but is now 'dead'.

■ BELOW: The blinding disc of the Sun rises over the wall of a scarp. We are standing at its foot, among the many huge blocks of rubble which litter the plain as a result of landslides.

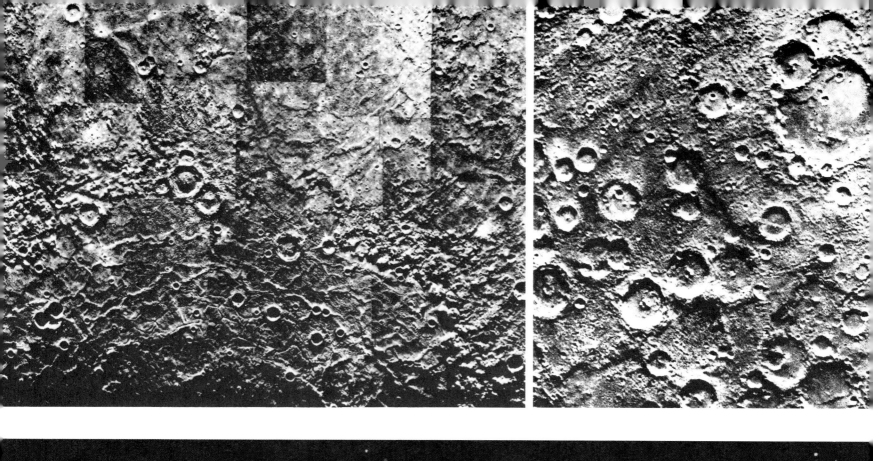

# Venus

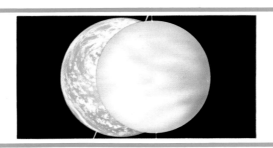

- Mean diameter: 12,104km.
- Escape velocity: 10.36km/s.
- Rotation period: 243 days (retrograde).
- Sidereal period: 224.7 days.
- Mean distance from Sun: 108,200,000km.
- Inclination of orbit: 3.39°.
- Inclination of equator to orbit: −2°.
- Mean density: 5.2.
- Surface temperature: 485°C max.

*❛"Look!" he said, his voice full of excitement. "Just to the left of that black mark! Tell me what you see." He handed over the glasses and it was Coleman's turn to stare. "Well I'm damned," he said at length. "You were right. There are rivers on Venus."❜*

Before Eden ARTHUR C. CLARKE (1961)

Apart from the Sun the brightest object in the sky of Mercury is the planet Venus. This is hardly surprising since firstly it is so close, and secondly it is intrinsically a bright object because of its high reflectivity or albedo. It reflects some 75 per cent of the light falling on it, compared with 40 per cent for Earth and about 6 per cent for Mercury and the Moon. This bright, white 'star', which is often the first to be visible when the Sun has set, is clearly visible in daylight if one knows exactly where to look.

The orbit of Venus is almost circular, 108,200,000km from the Sun. The size of the planet makes it very nearly Earth's twin – 12,104km to Earth's 12,756. Telescopically, its size varies considerably. When it emerges from the glare of the Sun after being at superior conjunction (see diagram on page 47, which applies equally to Venus) it is a small almost full disc. Each evening its apparent distance or 'elongation' from the Sun becomes greater, and it sets later after the Sun, until its greatest elongation, about 47° east of the Sun, when it appears in half-phase (dichotomy) and has more than doubled its apparent size.

From that point it begins to move back toward the Sun ever more rapidly, swelling to a narrow sickle six times as large as when full, until it is lost in the Sun's rays (inferior conjunction).

From then on the 'evening star' – the 'Hesperus' of Homer's Iliad – becomes 'Phosphorus', the 'morning star'. This is because it has passed in front of the Sun and its next appearance, a few weeks later, is before sunrise. Each morning it rises earlier until it reaches its greatest elongation west of the Sun, and back to superior conjunction.

An interesting phenomenon that occurs near inferior conjunction is the Ashen Light. This takes the form of a faint glow, which has been described as 'brownish', 'coppery' and 'greenish', on the planet's dark side. Venus, like Mercury, has no satellite, so this cannot be an equivalent to Earthlight on the Moon, and it has puzzled many observers. It has been ascribed to volcanic activity, airglow or aurorae. The recent discovery that Venus has no measurable magnetic field makes the latter unlikely, but airglow (caused by ionised molecules re-combining), remains a possibility. It is certainly not merely a contrast effect; neither, probably, is the oddity some observers claim to have seen: the night hemisphere standing out *darker* than the background sky. This may sound impossible, since the dark side of a planet cannot be darker than the black of space. But, when Venus is within 15° of the Sun its night side *could* be silhouetted against the Sun's outer or F-corona (even though daylight normally washes this out).

---

'Shining in the late dusk or early dawn, it [Venus] resembles a floating lamp, and when visible near Christmas time it never fails to produce a large number of enquiries from people who believe it to be a return of the biblical "Star of Bethlehem."'
*The Observer's Book of Astronomy*
Patrick Moore (1962)

---

When Venus turns its back on us a transit across the face of the Sún can occur, but only when the planet is at inferior conjunction and at a node. Since the orbit of Venus is inclined to the ecliptic by only 3° 24' this is quite rare (Mercury's orbit is tilted by 7°, so transits take place every few years). The first transit ever to be scientifically observed was seen by only two people, on December 4, 1639, one of whom was the man who predicted it, Jeremiah Horrocks. The previous one had occurred, unobserved, in December 1631; transits always occur in either December or June, and go in 'pairs'. That of 1639 was followed by four: June 1761 and June 1769, then December 1874 and December 1882. The next are due on June 8, 2004, and June 6, 2012.

The only phenomena of real interest during a transit are the 'ring of light' which surrounds Venus as it is about to touch the Sun's disc, and the 'black drop' which appears to be drawn after the planet when it enters the Sun's disc fully (see the diagram opposite).

When the planet is illuminated, the presence of an atmosphere is revealed by its cloudy nature. The surface of the planet cannot be seen at all, merely the brilliant yellowish-white of sunlight reflected from pearly cloud-layers. Because any faint markings are transitory and cannot be related to a rotation period, it was impossible to do more than guess this; even the axial inclination of Venus was in doubt. Cassini proposed a rotational period of 23 days in 1666, but changed it to 23 *hours* 16 minutes in 1667; Schiaparelli thought that Venus, like Mercury, had a synchronous rotation which would thus be 224.7 days – Venus' 'year'. But measurements showed that the temperature of the night side was too high for it never to receive any sunlight. By the 1950s a rotational period of between 15 and 30 days seemed the most probable.

But what attracted most interest was not the rotation of Venus, or markings on the clouds, but what lay beneath them. The most attractive idea, first put forward in 1918 by a Swedish chemist, S. A. Arrhenius, was that Venus, being almost Earth's twin in size but nearer to the Sun and thus hotter, was at a stage of evolution equivalent to our Carboniferous period, with steaming swamps and jungles. It did not take some writers long to add dinosaurs or other monsters.

The abundance of water accounted for the dense clouds; various theories were advanced to explain the lack of spectroscopic evidence for water. The main alternative theory, introduced in America a few years later, was that the surface of Venus is a 'dust-bowl', with clouds of yellow dust stirred up by

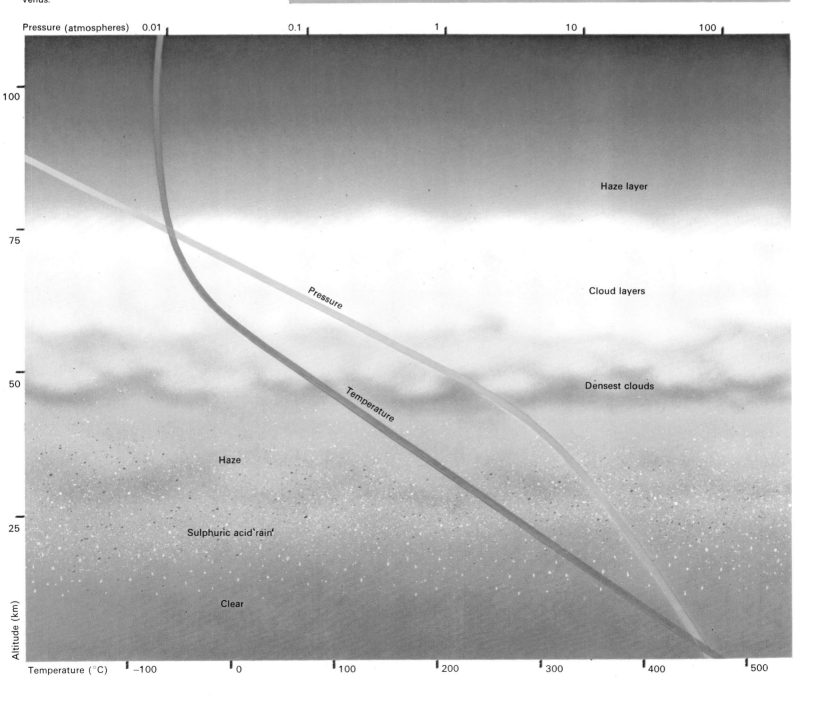

■ TOP: Sequence during a transit of Venus across the face of the Sun. The bright ring and the 'black drop' both prove the existence of an atmosphere. ■ INSET: The Ashen Light.
■ RIGHT: Apparent size of Venus at different phases. The actual date of dichotomy, or half-phase, usually appears to vary from that predicted. Venus makes 13 orbits in 8 of our years, which is also the interval from one inferior conjunction to the next, within a few days, so this can only occur around one of five dates. Venus also seems to exhibit spin-orbit coupling, rotating almost exactly four times between each inferior conjunction; so, like Mercury, it may present the same side to Earth each time.
■ BELOW: Temperature and pressure in the atmosphere of Venus.

Near inferior conjunction

Brightest crescent

Dichotomy

Gibbous

Superior conjunction

'Black drop'

Pressure (atmospheres)   0.01   0.1   1   10   100

Haze layer

Cloud layers

Densest clouds

Pressure

Temperature

Haze

Sulphuric acid 'rain'

Clear

Altitude (km)

100

75

50

25

Temperature (°C)   −100   0   100   200   300   400   500

strong winds and etching any rocks into fantastic, surrealistic sculptures. Many lurid 'artist's impressions' were made even up to the early 1950s (and I must plead 'guilty' – though I was careful to offer both alternatives).

The spectroscope did detect carbon dioxide in 1932. It was the 1960s before a second compound was detected – carbon monoxide. Water remained virtually absent, oxygen completely so.

In 1942 Rupert Wildt, in America, suggested that the temperature at local noon was 'probably as high as that of boiling water'. He showed that a greenhouse effect would arise in an atmosphere rich in carbon dioxide, due to the property of that gas in absorbing infrared radiation which would otherwise be re-radiated into space once an equilibrium was reached between the temperature of the surface and the radiation arriving. The absorbed infra-red heats up the atmosphere itself still further until the excess escapes from the top.

In 1967 Mariner 5 relayed information about the enormous amounts of carbon dioxide (97 per cent of the atmosphere) on the planet, together with information on cloud heights. James Pollack and Carl Sagan in the USA calculated the greenhouse effect this would produce and arrived at a temperature of 450°C. Mariner 5 was not the first US probe to encounter Venus. The first was Mariner 2 in 1962, and this reported a pressure of at least 75 atmospheres (Earth's at sea level being given a value of '1') and temperature of 650K. It also showed no magnetic field. Most important, perhaps, it confirmed radar observations made in the same year that Venus rotates in 243 days – with a backward, east-to-west or retrograde motion.

In 1968 radio and radar measurements gave a temperature of 477°C (750K), and a pressure of 90 atmospheres. Then, in 1969, two Soviet probes, Venera ('Venus') 5 and 6, landed on the surface and confirmed the radio and radar measurements. Previous USSR Venera probes had met with partial success: Venera 1 in 1961, Venera 2 and 3 in 1965 – the latter being the first crash landing – all suffered from communications failure. Venera 4, in 1967, descended by parachute on the night side and sent back information on the atmosphere down to about 25km, and Venera 5 also made a night landing in 1969, as did Venera 7 – which made the first *soft* landing – in 1970.

Venus put up a strong defence. In 1972 two US scientists, A. T. Young and S. Gill, independently found from polarisation data that the clouds are composed of sulphuric acid droplets. In the same year Venera 8 landed, and from measurements of the radioactive constituents of rocks on the surface concluded that the interior of Venus is differentiated. The core of Venus is in fact believed to be only slightly smaller than Earth's, although the lithosphere is different. The lack of a magnetic field is presumably due to the slow rotation of Venus – though these conclusions about the interior are drawn from only the one experiment on Venera 8.

## Beneath the Veils

In 1973 Venus was scanned from Earth by radar, and vast shallow craters were revealed on the surface. In February of 1974 Mariner 10 made its fly-by on its way to Mercury, taking some 3,000 photographs. The clouds were discovered to have a swirling, spiral structure, and the circulation of the atmosphere was studied, along with its chemical composition. Mariner 10 also took a series of photographs of markings on the cloud tops which showed them to have a retrograde rotation of four days, much faster than the planet's rotation.

The most visually interesting, and in some ways surprising, results came from Veneras 9 and 10 in 1975. These both went into orbit round Venus to become artificial satellites, but released lander capsules, both of which transmitted pictures of the surface. At one site remarkably sharp-edged rocks are visible, and at the other smoother, eroded rocks and boulders, both on a flat plain with a 'horizon' up to 300m away. The brightness of the light penetrating the clouds was also unforeseen. Winds at the surface were light, but at an altitude of about 40km there seem to be jet streams with speeds of up to 185km/hr probably accounting for the four-day rotation of upper clouds.

In 1977 the now upgraded 300m radio telescope at Arecibo produced radar images that showed craters and large volcanoes. Then in 1978 both Russian and American probes increased our knowledge about Venus considerably. Each time the US Pioneer Venus 1 passes below 4,700km of the surface during its orbit of the planet, a radar altimeter measures its exact height above the surface, and this is later rendered by computer into a coloured map showing various elevations. These maps complement those made by Arecibo, which can resolve details down to 10–20km across. The Pioneer orbiter measurements are accurate to within 0.2km.

Thus a picture has been built up of the Venusian 'continents' – which turn out to be merely raised areas on the vast, cratered plain which covers 50 per cent of the surface. One surprise from Venera 8 was that the lowlands seem to consist of a material very close to terrestrial granite, rather than basalt as expected.

One continental mass, located at about 65° north latitude, is comparable in size with Australia and was named Ishtar Terra. On this is a vast, pear-shaped plateau, Lakshmi Planum, with two great mountain ranges to its northeast and west. The roughest area of Venus is at the eastern end of this, and is called Maxwell Montes, nearly 11,000m above the plain – almost 2km higher than Mount Everest. Another, less mountainous land-mass, about 5° south of the equator and half as large as Africa, is called Aphrodite Terra, after the goddess of love. There is a huge rift valley at the eastern end of Aphrodite Terra, which drops 2,900m below the plain, with a smaller but deeper rift alongside. About 30° north of the equator rises a highland region, Beta Regio, apparently consisting of two massive shield volcanoes, of the type that comprise the Hawaiian Islands, but even larger – 4,000m above the plains.

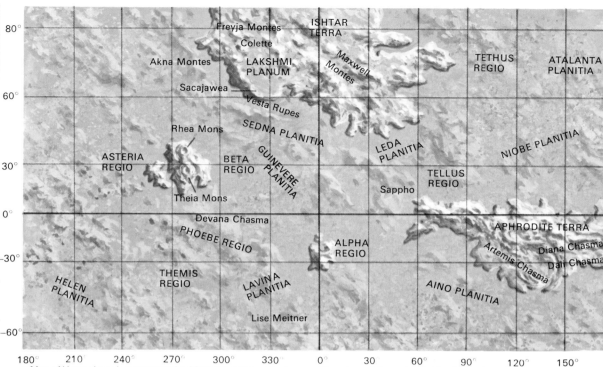

■ Map of Venus, based on computerised Pioneer-orbiter radar images. The edges of the major highland areas have been outlined more sharply for clarity. Prior to Pioneer, less than one per cent of Venus' topography had been measured, by Earth-based radar. There are many signs of volcanic activity, including lava flows; there may even have been some limited plate tectonic activity before the process was aborted. However, radar definition is not yet good enough to distinguish between some volcanic and impact craters (the latter, unusually, being found in the low areas, not highland).

■ ABOVE: Mariner 10 sent back these three ultraviolet images at seven-hour intervals on 7 February 1974, two days after it flew past Venus. The 1,000km-wide cloud feature indicated by the arrow reveals that the atmosphere rotates once every four days.

■ LEFT: Panorama of the surface of Venus, taken by the cameras of Venera 9 in October 1975. The sharp-edged rocks are 3-400mm across, and suggest that they were either formed recently (by volcanism) or exposed by faulting. At the bottom is part of the spacecraft; the arrow indicates a density meter lying on the surface.

■ RIGHT: Image of Venus made by Pioneer Venus orbiter on 14 January, 1978. The clouds near the poles are much thicker, and brighter, than during the Mariner 10 fly-by in 1974 (above). The dark 'Y' feature can be seen clearly, though it was much stronger a month earlier when Pioneer arrived.

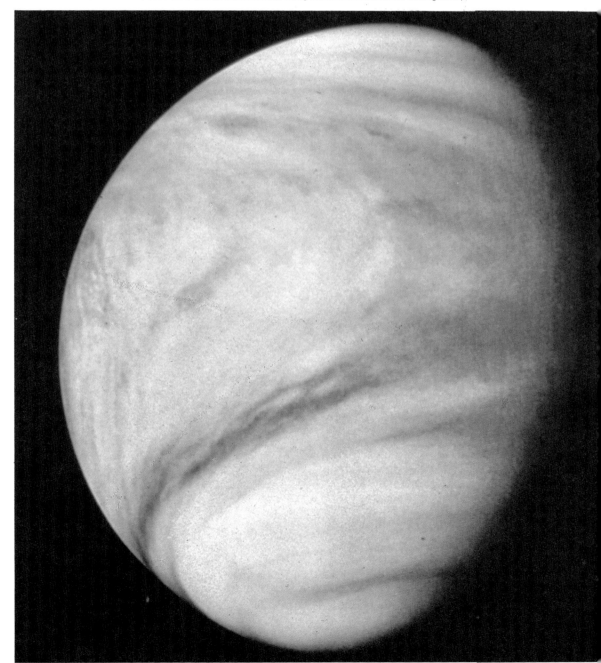

The other Pioneer to Venus was a 'multi-probe' mission. Pioneer Venus 2 carried four probes – three small, one larger – which were ejected from the transporter 'bus' about 20 days before it arrived at Venus and itself transmitted back information on the upper atmosphere before burning up like a meteor. The large, 1.5m probe parachuted down to about 44km then fell freely to a point where the equator crossed the terminator. The three smaller probes free-fell into the atmosphere some 11,000km apart on both day and night sides.

The four probes measured the pressure and temperature of the atmosphere, size and density of particles in the clouds, and infra-red sensors measured how much heat absorbed by the atmosphere comes from the Sun and how much from the planet's surface. They entered the atmosphere on December 9, 1978, four days after Pioneer Venus 1 took up its position as an observing station in orbit. Amongst the surprises found by the atmospheric probes was the news that Venus possesses hundreds of times more primordial argon and neon than Earth (though the proportion of each gas is about the same). This has caused many cosmogonists to revise their theories about the formation of the Solar System since, as we saw, the hot inner parts of the system *should* have been low in these elements. This suggests that there may have been a steady increase in density of *all* materials further in towards the protosun.

The probes confirmed that the 'greenhouse effect' is the major contributor in heating Venus' atmosphere. 25 per cent of the total sunlight falling on the planet is absorbed, but of the radiation attempting to leave a large part is trapped – not just by the carbon dioxide, dense though it is, but by *water vapour*: a surprising 0.1 to 0.4 per cent of water vapour was found in the atmosphere. The probes also found large particles of sulphur, both liquid and solid, which contribute to the heat trap.

The clouds are divided into three layers. Nearest the surface, from an altitude of about 48km, is a dense 5km-thick layer of opaque clouds, which appear to be made up of solid and liquid sulphur particles with a little sulphuric acid, at a temperature of 200°C. Above that, up to 56km, and 6km thick, is a layer of liquid sulphur and sulphuric acid particles; its temperature is down to 20°C. The upper layer, above 56km, is nearly 15km thick and may consist of sulphuric acid droplets only, at 13°C. There is also a high layer of haze, which is thicker over the pole and could contain water vapour or ice crystals, similar to Earth's cirrus clouds. Beneath the lowest clouds, too, there is a haze – probably of sulphuric acid particles – which thins out to give 'clear air' at 30km; there is also a 'rain' of sulphuric acid that vaporises as it falls into higher-temperature layers. The pressure and temperature are remarkably uniform at all latitudes, both day and night.

Two of the probes fell on the night side and provided most of the surprises. At an altitude of about 15km a strange glow was detected, which increased as the probes approached the surface. This may be due to high temperature chemical reactions in the hellish, sulphurous brew near the surface. Some form of electrical activity has also been suggested, and this is where the evidence from the Soviet Veneras 11 and 12 becomes interesting. These fly-by missions also released landing capsules, which reported continuous lightning flashes from 32km down to 2km altitude. The discharges occur at a fantastic 25 per *second* – too rapid to be separated by the human eye. The Venera probes also reported loud acoustic shocks (82 decibels) – thunder? The Pioneer orbiter also observed lightning discharges during every pass across the night side – which offers a possible clue to the puzzle of the Ashen Light.

Near the north pole the orbiter also found a depression in the clouds, and an actual 'hole', 1,050km across, where there are very few or even no clouds. This appears to be a permanent feature of the atmospheric circulation cells, and there could well be a down-flow of atmosphere at this point.

The atmosphere interacts with the solar wind in unexpected ways, too, in view of the fact that Venus has no magnetic field. Within Venus' ionosphere there are unique 'magnetic ropes', encompassing the whole planet, as though the interplanetary magnetic field were wrapped and twisted around the planet. It would seem that, with no magnetosphere to deflect it, the solar wind bombards the top of the ionosphere directly. Most of it flows around the planet, but the solar magnetic field builds up at the boundary and diffuses through the ionosphere in these curious, contorted field-lines. Another theory suggests that the electric current flows within the solar wind induce magnetic field 'spikes' in the ionosphere, but much research has yet to be done. Venus has had to yield some of her secrets from behind her cloudy veil – but many remain.

■ VAP—Venus Atmosphere Probe (not an actual project). A hydrogen-filled balloon (with no fear of explosion in a non-oxygen atmosphere) may be the best method to explore Venus. Signals will be relayed by an Orbiter.

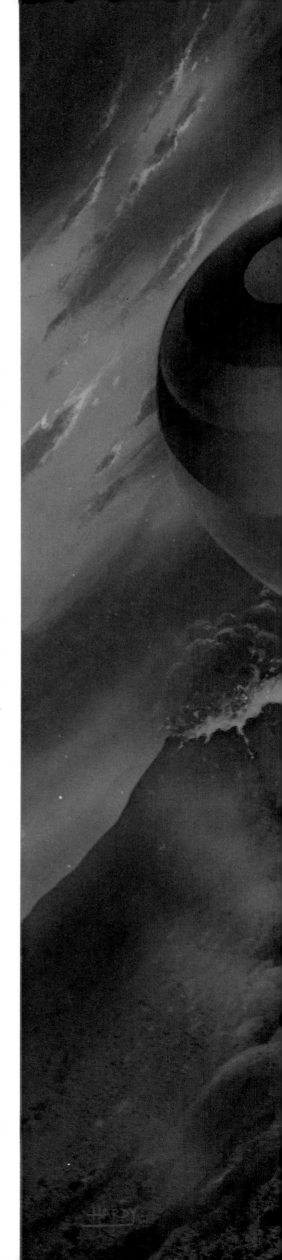

# Mars

- Mean diameter: 6,790km.
- Escape velocity: 5.0km/s.
- Rotation period: 24hr 37min 23sec.
- Sidereal period: 686.9 days.
- Mean distance from Sun: 227,940,000km.
- Inclination of orbit: 1.85°.
- Inclination of equator to orbit: 23.98°.
- Mean density: 3.9.
- Surface temperature: day max. 20°C; night min. −80°C.

*‘ There are on this planet, traversing the continents, long dark lines which may be designated as* canali, *though we do not yet know what they are.’*

GIOVANNI SCHIAPERELLI (1879)

Mars has long been known as 'the red planet' because of its characteristic ruddy glow in the night sky. The great Galileo first turned his telescope on this namesake of the Roman god of war around 1610. His observations soon revealed that its disc was not always a perfect circle. In fact its phase is never less than that of the Moon when a few days from full, because Mars' orbit is farther out from the Sun than Earth's. It is the closest planet to us, after Venus.

The first person to attempt a sketch of Mars' surface (so far as we know) was Huygens in 1659. His drawing shows a rather triangular area now known as Syrtis Major. It was soon realised that the markings were permanent and that we were looking at a solid surface. Even so, it was not until the space age arrived that we realised how wrong many of our notions about Mars had been!

The first piece of useful information about the markings was that they are carried around the planet in a west-to-east rotation as Mars rotates on its axis. So the Sun rises in the east and sets in the west on Mars, just as it does on Earth. In fact, the rotation period of Mars is very close in length to ours – 24 hours 37 minutes. Cassini made it 24 hours 40 minutes in 1666, which was close. He also observed brilliant white patches at the upper and lower edges of the limb, which his nephew G. F. Maraldi explained in 1719 as polar caps. Sir William Herschel noted that the dark areas, while permanent, did vary in appearance as if obscured by clouds, and that the polar caps changed size according to the Martian seasons, and in 1784 attributed this to melting snow.

The axial tilt of Mars is about 25°, less than two degrees more than Earth's, giving similar seasons. It is hardly more than half the size of Earth, its diameter being just over 6,790km, and its surface area is only 28 per cent of Earth's, its mass 0.1 per cent on modern estimates. But even so, in the early days of observing it was obvious that here was

another Earth, with seasons, polar ice and snow, continents or deserts (as the red tracts became known) and even seas (which were named 'maria', just as on the Moon), and clouds.

The orbit of Mars takes it 206 million km from the Sun at perihelion, 248 million at aphelion, and its revolution period or year is 687 days – about twice Earth's. When an 'inferior' planet (that is, one inside Earth's orbit – 'superior' is outside Earth's orbit) is in line with the Sun as seen from Earth it is said to be at conjunction. A superior planet is at opposition when it is in line with the Sun and both planets are on the same side of their orbits. It can then be seen all night.

The orbit of Mars is more elliptical than Earth's; it can approach to within 55.5 million kilometres, but it can be as far as 101 million kilometres, which makes a considerable difference to the size of the disc to Earth-bound observers. Obviously the best oppositions are when Mars is at perihelion, but such oppositions occur only every 15–16 years (the next will be in September, 1988), although, as the diagram shows, the two planets 'get together' every 780 days.

'That Mars is inhabited by beings of some sort or other we may consider as certain as it is uncertain what those beings may be ... to talk of Martian beings is not to talk of Martian men ...'
*Mars as an Abode of Life*
Percival Lowell (1908)

The first real map of Mars appeared in 1840, and laid down lines of latitude and longitude like those on our terrestrial maps. The geography of Mars is sometimes called 'areography', from 'Ares', Greek for Mars. Progressively better maps followed in 1869, 1872, and in 1877. Another favourable opposition occurred in 1877 and the director of Milan Observatory, Giovanni Schiaparelli, decided to make a 'trigonometrical survey' of the surface of Mars in order

to determine accurately the latitudes and longitudes of various features. While observing he found a number of new features, including some dusky streaks crossing the reddish 'continents'. He called these *canali*, ('channels'). Thus started a chain of events which has coloured our whole view of Mars, in fact and fiction – at least until the mid 1960s.

Most astronomers met Schiaparelli's claim with scepticism, since no-one else had reported seeing the 'canals', as many took the Italian word to mean. Undeterred, Schiaparelli discovered even more *canali* in 1879, and by 1881 they seemed more numerous, straighter, narrower and more geometrical than ever. There were even double lines where previously he had recorded only one.

The next important discovery came from Professor W. H. Pickering, who found canals not only on the continents but on one of the seas, the Mare Erythraeum. Then in 1894 Percival Lowell announced from his observatory at Flagstaff, Arizona that *all* the seas were crossed by canals. Clearly the 'seas' where *not* seas. Later in the same year Lowell observed changes in the dark areas, of an obviously seasonal nature. The dark regions must consist of vegetation, which grew and changed colour as the polar caps melted with the Martian spring. The canals, said Lowell, really were artificial water-courses contructed by intelligent Martians to carry water from the poles to irrigate their drying (and dying) planet. What we saw through our telescopes were wide strips of vegetation alongside the canals. The canals themselves were probably enclosed to prevent evaporation in the thin air – for it was by then known that Mars had an atmosphere, whose pressure was believed to be between a quarter and one-sixth of Earth's. Life, then, was a possibility if there was enough oxygen.

If one canal was insufficient a second

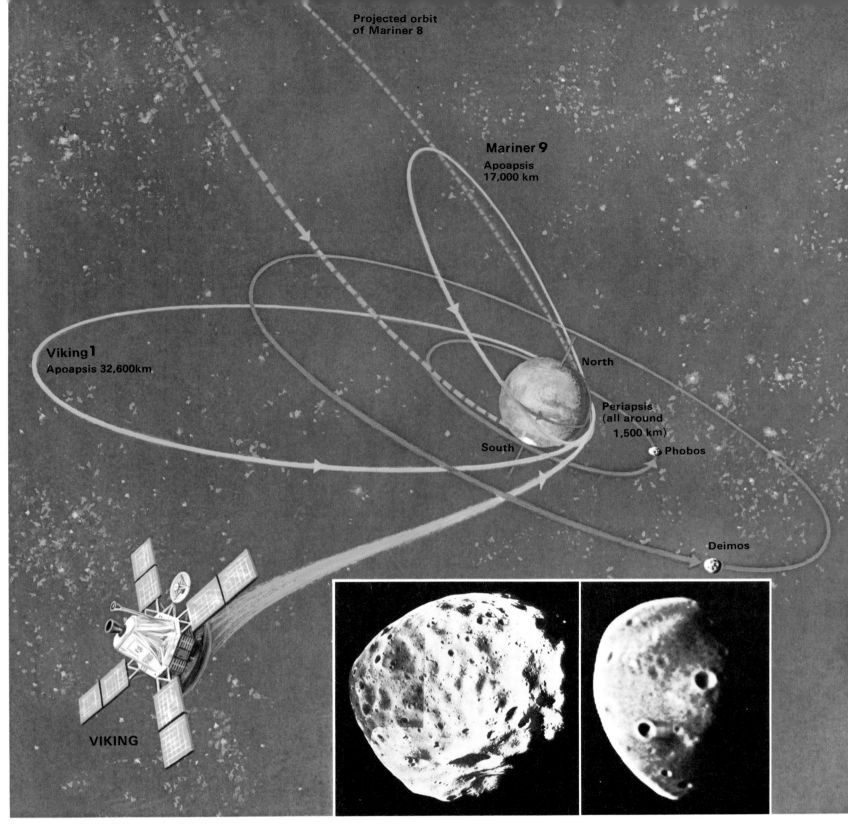

Projected orbit
of Mariner 8

Mariner 9
Apoapsis
17,000 km

Viking 1
Apoapsis 32,600km

North

Periapsis
(all around
1,500 km)

South

Phobos

Deimos

VIKING

■ ABOVE: Mars orbiters. The orbits of the two natural satellites are shown, to scale with Mars (though the moons themselves are enlarged), together with those of Viking 1 and Mariner 9. The dashed line shows the intended orbit of Mariner 8, which went off course soon after launch on 9 May 1971. The periapsis (closest point) of Viking 1's orbit passed over latitude 22°N in order to allow the lander to touch down at Chryse Planitia. Viking 2 had to pass over 48°N, but its orbit is omitted here for clarity. Mariner 9 had a high inclination of 65°. ■ INSET, ABOVE: Images of Phobos (left) and Deimos returned by Viking-orbiter 1. Both reveal heavily cratered and presumably very old surfaces. The smallest craters visible here on Phobos are 20m across, and on Deimos 100m. ■ BELOW: Oppositions of Mars and Earth. Because of the ellipticity of Mars's orbit, the next favourable opposition can be seen to take place in September 1988.

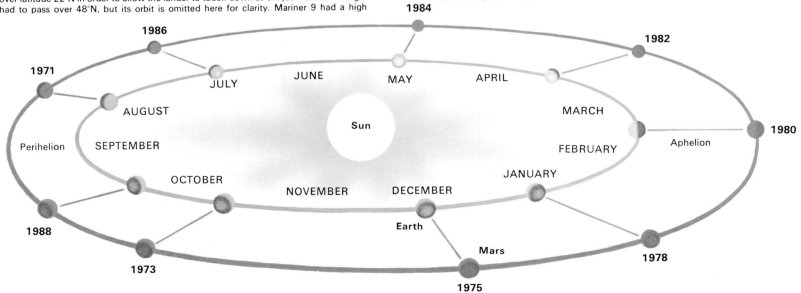

1984

1986

1971

1982

JULY

JUNE

MAY

APRIL

AUGUST

MARCH

Sun

1980

Perihelion

SEPTEMBER

FEBRUARY

Aphelion

JANUARY

OCTOBER

NOVEMBER

DECEMBER

1988

Earth

1978

Mars

1973

1975

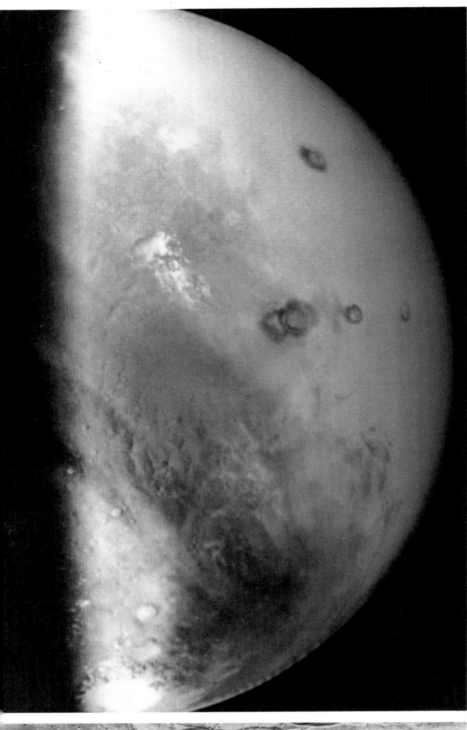

■ UPPER LEFT: The first Viking 1 colour photograph of Mars, taken from 560,000km, showed the four giant volcanoes in Tharsis, cloud near the poles, and ice fogs in the Argyre basin in the south.

■ UPPER RIGHT: A colour picture by Viking 2 on September 25, 1977 of late winter frost ($H_2O/CO_2$) on the red soil. In the foreground are parts of the lander, including the test charts used to calibrate colours. The colour TV picture was built up by scanning the scene through three filters—red, green and blue. A colour print was made on Earth by exposing conventional colour film to red, green and blue laser light, using the same scan pattern.

■ LOWER LEFT: Part of the incredible Valles Marineris, which stretches almost a third of the way round the planet, in a Viking 1 photomosaic taken from 4,200km.

■ LOWER CENTRE AND RIGHT: On 8 October, 1976 Viking 2's collector arm pushed a rock several centimetres, to collect a sample of soil from beneath it. It was thought that any lifeforms might shelter there from the Sun's intense ultra-violet radiation.

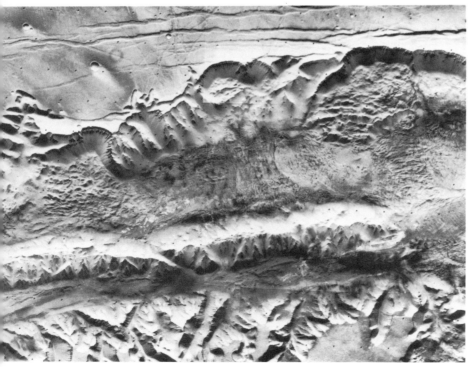

canal would be opened, explaining the doubling or 'gemination' of canals. Huge pumping stations would be placed at intervals, perhaps at the dark spots found at the intersections of several canals, which he called 'oases'.

But although other astronomers claimed to see canals, many, including Antoniadi, could see nothing so sharp – merely streaks. The canals have in fact never been photographed, and it does seem that the tendency of the human eye – or brain – to seek patterns may well be responsible.

For there are no canals on Mars. The detailed coverage of the planet by space probes since 1965 has found no trace of them. The whole conflagration of ideas, sparked off by Schiaparelli, fed with fuel by Lowell and fanned into flame by writers such as H. G. Wells around the turn of the century, was built of dream-stuff. Wells' *The War of the Worlds,* published as a book in 1898, probably contained the first appearance of the concept of Martians having aggressive tendencies toward the inhabitants of Earth. But the flames were fed by the authors of many, many stories, novels and films, all of which at least helped to foster public interest in science, astronomy and space travel.

The first spacecraft from Earth to reach Mars (apart from Mars 1, a USSR probe which lost contact in March 1963) skimmed past on 15 July 1965. Mariner 4 transmitted 21 historic close-up pictures from 9,846km. There were no canals, but something which few people had expected: lots and lots of craters.

The surface of Mars looked disappointingly Moon-like in those black-and-white Mariner 4 photographs, which showed craters down to 3km across. The next US probe, Mariner 6, passed 3,436km from Mars on 31 July 1969. It returned 50 pictures during approach and 26 as it flew over the equator. Mariner 7 passed on 5 August 1969 at a distance of 3,200km and sent back 93 photographs, with an additional 33 as it passed near the south pole, where the lowest temperature was found to be −153°C. Mariners 6 and 7 both reported more craters, with details down to 300m; Mariner 6 discovered two new types of terrain, one smooth and the other chaotic. It also confirmed that the atmospheric pressure at the surface is *very* low – less than 1 per cent of that on Earth. (It was later put at about 0.7 per cent.)

## The Race to Mars

In May 1971 two Russian probes, Mars 2 and 3, landed, but communications seem to have failed at that point. The first artificial satellite of Mars was Mariner 9. After a 15-minute braking burn to ensure that Mars' gravity net would snare it, the craft was placed into an orbit that initially took it from 1,398 to 17,916km from the surface, on 13 November 1971.

During the approach, scientists at the Jet Propulsion Laboratory at Pasadena had been somewhat frustrated by huge dust clouds – observed from Earth as 'yellow clouds' – which obscured much of the planet. A 200km spot visible near the equator had mysterious long streaks extending northward from it; three other vague spots were also visible.

When Mariner 9's transmitter was turned off on 27 October 1972, the spacecraft had made almost 700 orbits and sent back 7,329 photographs. Altogether, the whole of the surface of Mars had been surveyed. It also took photographs of the two tiny moons (which were named after the god of war's chariot horses); Phobos (Fear) proving to be an irregular, potato-shaped, crater-pitted lump of rock, 27 by 21 by 19km, while Deimos (Panic), similar in appearance, is about 15 by 12 by 11km. It reported that water vapour does exist in fair quantities over the south pole cap. Mariner 7 had found only solid carbon dioxide ('dry ice') in the thin white deposits there. Carbon dioxide comprises 95 per cent of the atmosphere, with 2.7 per cent nitrogen and various other gases.

The *average* surface temperature was found to be about −20°C (290K) at the equator; it can become as hot as 20–30°C at mid-summer, which is comparable with summer in England! There are 'hot spots' on the night hemisphere with temperatures 10–25°C higher than the area around them. On the day side, the dark areas are warmer than the light deserts (dark surfaces absorb more infra-red radiation than light materials).

As Mariner 9 arrived at the start of the northern spring it was hoped to confirm or otherwise the 'wave of darkening' which observers long claimed to

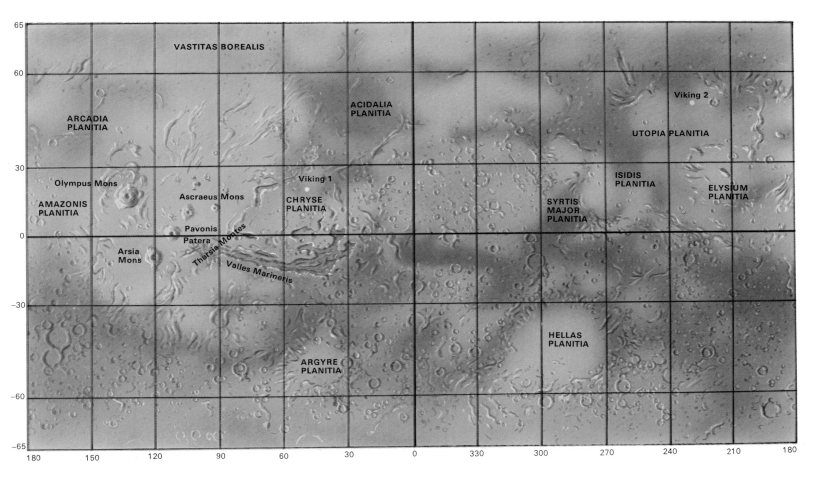

have seen spreading from the polar cap as it shrank. In the event, the dust storms obliterated any such surface change, but it seems that the 'seasonal changes' are due to no more than a redistribution of dust by winds and summer dust storms. The changes of colour of the dark regions, from blue-green in spring to chocolate-brown in autumn, must seemingly be put down to contrast effects and/or wishful thinking.

Mariner 9 reported that Mars, like the Earth, has an equatorial bulge; the diameter at the equator is 6,790km compared with a polar diameter of 6,746km. Like the Moon, Mars has concentrations of denser material, or 'mascons', beneath some parts of its surface, giving it an uneven gravitational field.

Its most interesting features, however were topographical. Once the dust storms had abated, a number of giant volcanic cones became visible. So too did a vast system of interlinked chasms or gorges, the largest of which – Valles Marineris – would dwarf America's Grand Canyon. Strange, sinuous channels looking remarkably like dried-up river beds were also seen and, in the polar regions, great sedimentary deposits.

During 1973 Russia launched a full-scale assault on Mars. Ironically, the 'red planet' repelled all would-be Soviet boarders. Mars 4 failed to go into orbit on 10 February 1974; Mars 5 did go into orbit on 12 February 1974, but transmitted information for only two weeks; Mars 6 lost contact while landing on 12 March 1974; and Mars 7 failed to land at all and sailed past on 9 March 1974. Some of the landers may have been caught in high winds during a dust storm, or overbalanced on a large boulder.

The US Viking 1 and 2 probes were launched on 20 August and 10 September 1975 respectively, and were an unqualified success. After several weeks' reconnaissance from orbit for the most suitable landing sites, each separated into an orbiter and a soft-lander, the latter using a parachute and retrorockets. Viking 1 landed on 20 July 1976 in Chryse Planitia, the 'Plains of Gold'. The Viking 2 lander touched down safely on 4 September 1976 in Utopia Planitia.

## Some Martian Surprises

The photographs taken by the Viking orbiters complemented and added to the information from Mariner 9. They found even more signs of the apparent past presence of water than had the Mariners, because the Vikings had high-resolution cameras, capable of considerable enlargement when received and 'enhanced' on Earth.

A group of four huge volcanoes is prominent in the Tharsis region, a little north of the equator, and there is another group in the Elysium region. Of these, the biggest – and probably the largest volcano anywhere in the Solar System – occurs in the first group. The IAU named it Olympus Mons. It stands some 25km above the desert, and is 600km across at its base. It is also interesting because, although no-one knew it at the time, it was seen from Earth. In 1879 Schiaparelli observed a small, brilliantly white spot about 20° north of the equator. He supposed it to be snow, and christened it Nix Olympica – 'snows of Olympus', home of the Greek gods. It is there that Olympus Mons stands. The giant cone was also the feature which first stood above the dust storm when Mariner 9 arrived; the streak was its 'wind shadow'.

Such spots, for there were others, puzzled early observers. Most knew that they could not in fact be snow, because they appeared in summer, at the warmest time. It was tentatively suggested that they may be due to crops growing! Isolated condensation clouds *do* however appear at equatorial latitudes. The air rises and cools over elevations such as the volcanic mounds and become brighter as the day progresses in summer, when there is a higher water vapour content in the atmosphere.

The misconceptions and preconceptions about Mars meant that much of what was eventually revealed came as a surprise. Because the atmosphere is so thin, just about everyone (artists included!) expected it to be dark blue, almost black. When Viking 1 landed and began to transmit its first TV scan

■ Photomosaic of the Nilosyrtis region of Mars. The strange humocky terrain resembles terrestrial features in which near-surface material flows very slowly due to the freezing and thawing of water frozen between its layers.

■ The crater Yuty, in the eastern Chryse area, exhibits a particularly well-defined ejecta blanket. Unlike the lunar rays, the blanket has a liquid appearance, like mud in which a stone has been dropped.

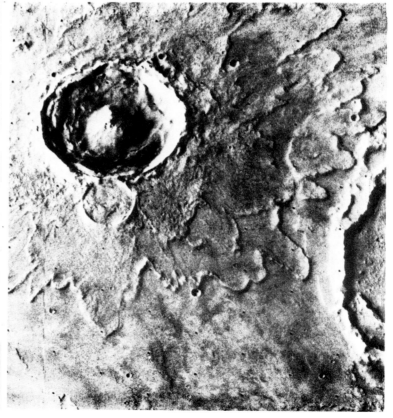

in colour, waiting scientists were worried by the unexpected reddish glow that bathed the whole scene, ground and sky alike. So the colour of the sky was adjusted to a neutral grey; but during photo-reproduction it took on an Earth-like blue tint, which is how it was released to the press. Later analysis showed the surprisingly bright, orange-pink colour of the sky to be accurate. The effect is caused by fine dust particles suspended in the thin air and scattering the Sun's light.

The Vikings' meteorology experiments reported winds of usually less than 17km/h, but gusts of over 50km/h. Much stronger winds must raise the dust storms, and wind-carried dust has 'weathered' the craters for billions of years, so that they look smoother and flatter than those on the Moon or Mercury. It also produces fields of dunes, and streaks from the leeward side of craters.

## Olympus Mons

Not everyone got it wrong. In 1954 D. B. McLaughlin, an American, wrote: "Quite possibly cones as large as Mounts Rainier and Shasta have been missed", when expounding his theory that volcanism played an active role on Mars, and that the dark markings were due not to vegetation but to volcanic ash. Even he did not expect a volcano three times as high as Mount Everest. He had calculated that cones the size of those US mountains would not be revealed by the shadows they would cast near the terminator. The lack of such shadows had long led astronomers to believe that Mars was flat and uninteresting.

It has been estimated that if it grew at terrestrial rates, Olympus Mons must have taken 10 million years to reach its present size by repeated eruptions of lava. Curiously, all the volcanic activity on Mars seems to have been limited to its northern hemisphere, while most of the flat-bottomed impact craters are found in the south. This indicates that the volcanic areas are considerably younger. Both the Tharsis area, on which Olympus Mons stands, and Elysium are structures known geologically (and now areologically) as domes – deformation of the crust, making it rise several kilometres above its surroundings. Several cones appear on both the Tharsis and Elysium domes, and they resemble the Hawaiian shield volcanoes (see page 52).

Also associated with volcanic features such as domes, cones and lava flows are faults and fractures in the crust. There is an abundance of evidence of these on Mars. A radial pattern of faults covers about a third of the planet; it is centred on Tharsis. But the most spectacular of these fractures is the colossal Valles Marineris (previously called Coprates Canyon). At its western end it forms a lattice-like network of grooves, then it extends like a great gash or scar for 4,000km across the skin of the planet before vanishing into low, chaotic terrain. It is nearly 200km across at its widest points, and 6km deep. (By comparison, our Grand Canyon is, in all, 445km long and 1.6km deep.)

No river excavated the Valles Marineris as the Colorado did the Grand Canyon. It is a deep rift caused by movements of the Martian crust. Yet even the Grand Canyon needed help from faulting in Earth's crust, caused by plate movements, before rivers could begin to deepen the cleft in the *dome*-shaped plateau through which the canyon runs. That slow erosion process started 10 million years ago, scouring rocks already 2 billion years old.

Some scientists believe that there was some plate tectonic activity at one time on Mars; hot convection currents in the mantle disrupted the crust and volcanoes began to erupt, flooding and smoothing ancient impact craters with seas of lava. But the process halted before the sort of wide-scale horizontal sliding and shifting of crustal plates that we find on Earth could take place. With less crustal movement a volcano could sit over one magma pocket for long enough to build up a huge cone – unlike the Hawaiian islands, which are the tips of a 'string' of cones forced up as the crust slid across a hot, rising plume in Earth's mantle.

## Flash Floods

Other experts doubt this theory, and believe the canyons to be due to the collapse of the surface as volcanoes emptied underground magma chambers, followed by long erosion processes, largely by the wind.

The present lack of water does not mean that it did not at some time play a part in shaping the surface of Mars. While the tectonic and/or volcanic activity continued it would have been accompanied by a considerable increase in atmospheric pressure, as outgassing took place. Gases, which included *water vapour* (as steam), gushed from fissures in the ground, and as they cooled in the atmosphere condensation would have occurred. It may have rained on Mars. 3.5 billion years ago 'flash floods' may have foamed over the red soil, churning it into mud as they carried rocks and debris before them. The sinuous channels, so reminiscent of aerial photographs of Earthly river beds, may actually have been racing torrents.

Or were they? American scientists J.

A. Cutts and K. R. Blasius have suggested that water was not responsible after all, but lava. Large amounts of fluid lava, they say, can dissolve the surface rock and cut channels, just as water does in ice and snow.

But there still seem to be several unique Martian features that can be explained more elegantly by the presence of water, or of ice. The 'chaotic terrain' on Mars, for instance, could be due to the withdrawal of sub-surface material followed by the melting of surface ice. The 'fretted terrain' consisting of odd, isolated hills or 'mesas' may be due to the melting of ice beneath the surface, causing landslips. Then there is the 'braided terrain', which looks very like deposits of silt in a channel after a flood has passed. And the mud-like 'splatter' surrounding the crater Yuty, due possibly to an impact that melted subsurface permafrost (ice in permanently frozen ground, as found in Arctic tundra). And the area on the Chryse plain over which a flash flood seems to have swept, leaving only a few craters high and dry.

Dr. Carl Sagan of Cornell University has suggested that even quite small amounts of water could account for the erosion we see on Mars – if the rivers had been protected by a thick layer of ice. They could even be flowing today. Another form of 'layered' or 'laminated' terrain, which also occurs in polar ice, suggests that periodic changes in climate may have occurred. How? Sagan and his colleagues calculated that if the temperature at the poles were increased by only 10–20 per cent, much of the carbon dioxide dry ice would vaporise. The total mass of the atmosphere would increase and it would carry more heat to the poles, causing *more* carbon dioxide to be released, and so on, creating a greenhouse effect and simultaneously increasing the water vapour content of the atmosphere. The global climate would change drastically – and may do so in the future.

## Long Lost Atmosphere

Certainly a massive atmosphere due to outgassing must have existed in the past. The amount of the heavy and chemically inert gas argon remaining on Mars from that time can be used as a guide to the amount of other gases that originally accompanied it but have either escaped into space, combined with elements in the soil (eg. iron, whose oxides and other compounds give Mars its red colour), or frozen out as permafrost. Viking's measurements of argon bear out the existence of an atmosphere up to half as dense as Earth's, long ago. Much of that original atmosphere is now thought to be locked

up in the surface rocks of Mars.

Contributory factors to a warming process at the poles that could release a dense atmosphere could be: an increase in solar radiation over millions of years; volcanism in ice-covered areas, as found in Iceland; the regular cyclic change in the inclination of Mars' axis, which 'nods' toward or away from the Sun every 100,000 and 1 million years; the reflectivity of the polar regions may change because of variations in the amount of material deposited by dust storms; or, as suggested by fellow-writer

Duncan Lunan, an impact by a carbonaceous chondrite asteroid might have scattered dark material in the atmosphere and over the poles, absorbing more infra-red radiation. (As it happens, the two moons of Mars are composed of very dark material.)

By an amazing coincidence, long before anything was known about the two satellites of Mars, Dean Swift wrote in *Gulliver's Travels* (1726) of his fictitious astronomers' discovery that Mars has two moons: "the innermost is distant from the centre of the primary

planet exactly 3 of its diameters, and the outermost 5; the former revolves in the space of 10 hours, and the latter in $21\frac{1}{2}$". It was not until 1877 that Asaph Hall discovered that Mars does indeed have two moons. Phobos is 9,000km from Mars, orbits in 7.65 hours, and appears less than half the size of our Moon. Deimos is 23,000km from Mars and revolves in 30.24 hours, looking like a bright star from Mars. Both are probably captured chondritic asteroids.

And what about those Martians? Life could perhaps have emerged in an early,

warmer epoch. But the Viking landers each carried five experiments to test for signs of carbon-based life. Two gave negative results. The three experiments that looked for the effects of metabolism, respiration and photosynthesis gave results which first looked positive, then merely ambiguous. For it was shown that the gases released could have been produced by purely chemical reactions. It seems that Martian soil is unusually active chemically. But we have so far examined only a *very* small patch of Mars.

■ Early-morning 'fog' – clouds of water-ice – forms in a canyon in the Noctis Labyrinthus plateau area of Mars. The outlines of the cliffs on the usually sharp horizon are softened as the fog spills over onto the plateau. It is probably caused when water, which condensed and froze during the previous afternoon on east-facing slopes of the canyon floor, is vaporised by the rays of the rising Sun. The inner moon, Phobos, is rising – in the west. This is because it orbits so much more rapidly than Mars rotates. It will set in the east 4.5 hours later, going through over half of its cycle of phases, and rise again in 11 hours.

■ RIGHT: Cross-section of Mars. Earth's core is only slightly larger, proportionately. The crust of Mars is, however, deficient in magnesium; in the northern hemisphere light, silicon-rich molten rock has flooded to the surface, forming vast lava-plains and building volcanic cones.

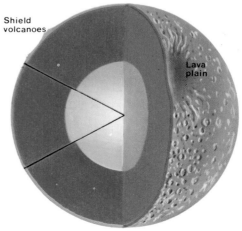

Shield volcanoes

Lava plain

# Cosmic Debris

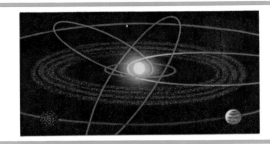

| Asteroid | Diameter | Mean distance from Sun (AU) | Orbital inclination |
|---|---|---|---|
| 1 Ceres | 1,000km | 2.76 | 11° |
| 2 Pallas | 600km | 2.77 | 35° |
| 3 Juno | 250km | 2.67 | 13° |
| 4 Vesta | 540km | 2.36 | 7° |

| Comet | Distance from Sun (AU) perihelion: | aphelion: | Orbital inclination | Period (years) |
|---|---|---|---|---|
| Encke | 0.3 | 4.1 | 12° | 3.3 |
| Halley | 0.6 | 35.3 | 162° | 76.0 |

**'*Inter Jovem et Martem planetam interposui.*
(Between Jupiter and Mars I put a planet.)'**

JOHANN KEPLER

On New Year's Eve of 1800–1801, Professor Giuseppe Piazzi, Director of Palermo Observatory in Sicily, was making star observations with the intention of correcting a mistake in a new star catalogue. He noticed a faint star where no star should have been. On subsequent nights it moved against the stellar background, and he thought that he had discovered a comet. He passed his measurements to Karl Gauss in Germany. Gauss, a great mathematician (though then a young man), quickly realised that the orbit of the mystery object was not elliptical, as a comet's should have been, but almost circular.

It seemed that Piazzi must have found a new planet – an addition to the seven planetary bodies then known. For Uranus had been discovered far beyond Saturn only 29 years earlier. Significantly, this new body was between the orbits of Mars and Jupiter, where a large 'gap' had long been known to exist.

That a 'gap' existed at all was known from what appears to be a fairly arbitrary manipulation of numbers according to a rule known as the Titius-Bode Law (or rather unfairly, as 'Bode's Law'). The distance between the Sun and the Earth is used and for measuring purposes is counted as one astronomical unit (AU). For the purpose of the Titius-Bode Law it is taken as 10. The formulator of this rule was a professor from Wittenburg, J. B. Titius, who inserted it as a footnote in a book. Another German, J. E. Bode, found this note and published it in 1772.

| Planet | Titius-Bode distance | Actual distance |
|---|---|---|
| Mercury | 4 | 3.9 |
| Venus | 7 | 7.2 |
| Earth | 10 | 10.0 |
| Mars | 16 | 15.2 |
| — | 28 | — |
| Jupiter | 52 | 52.0 |
| Saturn | 100 | 95.4 |

It goes like this: take the numbers 0, 3,

6, 12, and so on, doubling each time. Add 4 to each and you get the numbers in the centre column below. If we take these as distances from the Sun, with Earth as 10, it is remarkable how closely the other planets fit.

At the time of Titius and Bode no planets were known beyond Saturn. During the late 18th century astronomers searched avidly for the 'missing planet' between Mars and Jupiter. (Over 150 years earlier Kepler had suspected the same, and wrote the Latin lines quoted in the heading to this chapter.) When Uranus was discovered in 1781 it fitted neatly after Saturn: the Titius-Bode relationship gives 196, the actual distance is 191.8. Neptune did not fit when it was found, unfortunately, but Pluto is close, if taken as following Uranus, at 388 and 394.6.

'In the first year of Yuen-yen, in the 7th month, on the day sin-ouei [25 August] a comet was seen in the region of the sky known as Toung-tsing [μ Gemini]. It passed over [Gemini] proceeded from the Ho-su [Castor and Pollux] in a northerly direction and then into the group of Hien-youen [the head of Leo] and into the house of Thaiouei [tail of Leo] . . . On the 56th day it disappeared with the Blue Dragon [Scorpio]. Altogether the comet was observed for 63 days.'
The first description of Halley's comet in 12 B.C. from the Wen-hien-thung-khao encyclopaedia of the Chinese scholar Ma Tuan-lin.

There is as yet no theoretical explanation for the Titius-Bode 'law'. But to the astronomers of 1801, the discovery of a planet at 2.77AU, or 27.7 on the Titius-Bode scale, was vindication of the law.

Piazzi's discovery was named Ceres, but was found to be only about 700km in diameter (modern measurements give over 1,000km), so it was not large enough to qualify as a true planet. But there were more surprises in store. A *second* small planet was found near 2.8AU on 28 March 1802 by Dr. W. M. Olbers, an amateur astronomer who was comet-spotting at Bremen. It was called Pallas, and is 480km across. Olbers discovered another, Vesta, in 1807, but

not before the third, Juno, had been found in 1804 by another German astronomer, K. Harding.

The idea, first suggested by Olbers, was by now growing that since several small bodies had turned up instead of the expected major planet, then perhaps these small objects could be fragments of a planet that had disintegrated or exploded. It would therefore be logical to expect more, and the searchers pursued their quest diligently. None were found.

In 1830 a Prussian amateur observer, K. J. Hencke, began a search for more of these 'minor planets' or 'asteroids', as the four had become known. In 1845, 38 years after the discovery of Vesta, he came across Astraea. By now the new star charts that had been used in the discovery of Neptune had come into use, and new asteroids were found by other astronomers at an increasing rate. Hebe, Iris and Flora were all found in 1847; Metis followed in 1848 and Hygeia in 1849. But all were less than 160km across.

By 1870 a total of 109 asteroids had been assigned names and numbers, and their orbits had been computed; by 1890 there were 300. Long before then the names of classical goddesses had run out, and there was great controversy over new names. The twelfth asteroid was named Victoria in 1852 but after objections it was re-named Clio. Isabella and Angelina broke the classical tradition, and by the time No. 232 was reached the namers had started on countries: Russia, Germania, Italia, America; then cities, districts, colleges, astronomers' names.

Largely responsible for the asteroid 'boom' was a new method of detecting them, suggested by Dr. I. Roberts and implemented by Professor Max Wolf of Heidelberg in 1891: photography. An exposure time of one or two hours with a camera attached to a telescope fitted with a 'clock drive' (to eliminate the rotation of the Earth) ensures that stars appear as points of light, but any

■ ABOVE: In this 'long-focus' picture, the asteroid Hidalgo looms above Jupiter and its four Galilean satellites, only some 0.4AU away. Its huge orbit (the largest known) takes Hidalgo between 2.0 and 9.7AU from the Sun; and it is the only minor planet to pass within 1AU of Jupiter.

■ BELOW: The first four asteroids to be discovered, with the oddly shaped Eros and the Moon to the same scale. Several asteroids are in fact slightly larger than Juno, but estimates vary in all cases. The well-defined main zones are separated by 'Kirkwood Gaps', named after their discoverer in 1866.

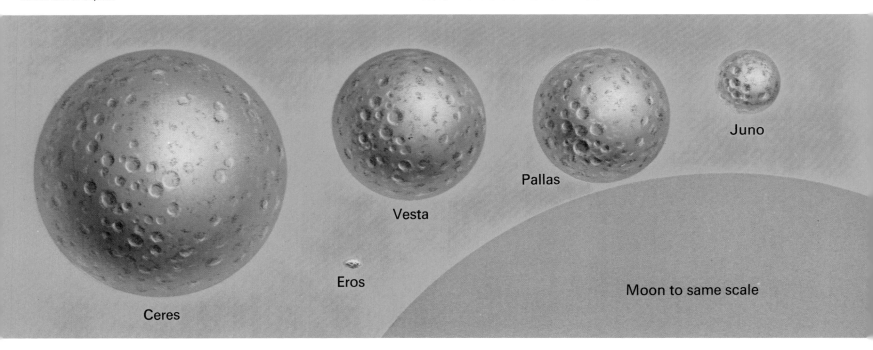

Juno

Pallas

Vesta

Eros

Moon to same scale

Ceres

asteroids show up as short lines or trails. Wolf ended up with a total of 582 asteroids by this method. Female names quickly ran out altogether, and a system of using the year of discovery followed by two letters was adopted in 1925, providing for two new asteroids to be added each *day*.

Most of these chunks of rock circle the Sun in well-defined zones between Mars and Jupiter. No. 433, Eros, was discovered in 1898. It proved to have an orbit which could bring it much closer to the Sun at perihelion; it could in fact approach Earth within 24 million km as it did in 1931 and 1975. Its shape is also odd, as it seems to look rather like a bread roll – 27km by 15km – as evidenced by fluctuations in its magnitude or brightness over a period of about 5 hours.

Several other 'Earth-grazers' were subsequently found. Amor can overtake the Earth at a distance of 16 million km but Hermes passed only 780,000km away in 1937 (it is only 1km across). Apollo can pass at some 3 million km and its orbit actually extends towards Venus. But Icarus can approach the Sun closer than Mercury – 28 million km from the solar furnace at perihelion, when it must glow red-hot (500°C) before retreating into the cool depths beyond Mars again. Hidalgo, at the other extreme, can almost graze the orbit of Saturn. Then there are two groups, known as the Trojans, which keep roughly 60° ahead and behind Jupiter forming an equilateral triangle with the giant planet, at the 'Lagrangian points' (after Comte Joseph Louis Lagrange) in its orbit. For some reason

there are twice as many in the 'fore' position than the 'aft'.

Some 2,000 of these flying mountains are now recorded. The total number may exceed 50,000, or even 100,000, plus many more too small to be observed from Earth.

There is, of course much speculation about their origin. Are they really the remnants of an exploded planet? This theory was popular for a long time, but had to be discarded because if they were all lumped together the asteroids would produce a body smaller, less massive than the Moon. A clue to their origin might be gained by examining their probable composition.

They fall into two main types: rather reddish bodies with a fairly high albedo and probably consisting of silicates intermixed with metals; and those of

■ TOP: In a still picture a comet and a meteor both appear to be shooting across the sky, but the meteor will actually have vanished in seconds, while the comet remains visible for many nights. **Inset**: The orbit of a short-period comet showing how the tail moves with respect to the Sun. The colours of the dust and plasma tails are exaggerated for clarity.

■ RIGHT: A shower of meteors appears to originate from a point known as a radiant. This is purely a perspective effect, comparable with parallel railway lines appearing to meet at the horizon.
■ FAR RIGHT: Why 'shooting stars' are more likely to be seen after midnight than before. Between 12.00pm and dawn the night side of Earth meets meteoroids 'head-on'; until then, they have to travel very rapidly to overtake Earth in its orbit.

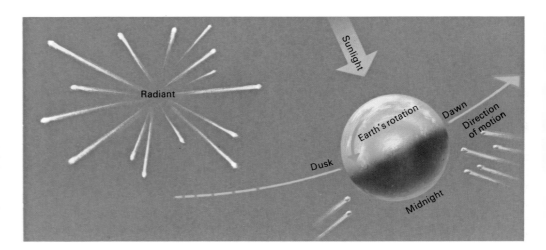

greyish colour with low reflectivity (3–9 per cent). We already know that carbonaceous chondrite meteorites contain some of the most primitive materials in the Solar System, minerals that formed as the primordial nebula cooled. The spectra of some 80 per cent of asteroids examined spectroscopically are found to resemble the spectra of this type of meteorite very closely, while 10 per cent correspond with the 'stony iron' type. The former underwent the least metamorphosis, contain a high proportion of carbon and much water in combination; the majority of the latter are found in the asteroid belts closest to Mars, and are in the 100–200km class. Most of the asteroids are probably highly cratered, as Mars' moons were found to be; as we saw, Phobos and Deimos may be no more than captured asteroids of the chondritic type – they are very dark in colour. However, their very circular orbits make some scientists doubt this. Some of the smaller satellites of Jupiter and Saturn are almost certainly asteroids snared by those planets' gravitational fields.

So, rather than being the remains of a planet, it is much more probable that the asteroids are the left-over debris that simply failed to accumulate material when the accretion processes in the original nebula took place. Perhaps Jupiter captured the lion's share of material in that vicinity, and grew fat as a result. Ceres, perhaps, had just started to grow to planetary size when the accretion processes petered out. (Incidentally, if all the other asteroids in the main belts were put together, they would make one Ceres-sized body.)

Clearly there is a strong connection between asteroids and meteoroids. The best interpretation is that the (generally larger) silicate/metal asteroids may be the remains of the metallic cores of large bodies in which heating and differentiation had started to take place, but which had their outer silicate mantles knocked off them by repeated violent collisions. The chondritic asteroids never grew large enough to become very hot inside; or those remaining *as* asteroids are the ones that were lucky enough not to have been involved in collisions which would have split them. The fragments of both types eventually scattered throughout the Solar System until they became attracted by the gravity of one planet or another.

Some meteorites do show fracture marks which indicate that they have been in high-velocity collisions, and there are other clues that they originated from either the interiors or the surfaces of larger parent bodies. In the case of iron meteorites the indications are that their parent bodies were some hundreds of kilometres across. If two asteroids collide at a high relative speed, both will be destroyed; their debris (if rocky, metamorphosed by the heat generated in the impact) is likely eventually to form craters on some larger body.

The formation of an impact crater is not quite so simple as it may appear. For one thing, the crater will be circular whether the meteorite hits 'head-on' or at an angle (except in the case of a really glancing collision, which can leave a groove). The body arrives at such a high speed – tens of kilometres a second – that the material into which it burrows cannot move out of the way quickly enough, so becomes highly compressed. In slowing down, the meteorite's potential and kinetic energy is converted into heat energy – enough to vaporise a large part of both meteorite and surface material. Assuming that the meteorite is large enough not to be completely vaporised, its solid kernel will come to rest perhaps kilometres below the surface in a seething pocket of incandescent, compressed rock, which immediately explodes violently upwards and outwards, creating the typical profile of a crater.

## 'Shooting Stars'

Some particles never reach a surface to make a crater. Many tonnes of meteoric debris enter our atmosphere every day. Much of this consists of micrometeorites – dust which drifts in the atmosphere until it reaches the ground. The meteors or 'shooting stars' which you can see on almost any clear night are usually no larger than a grain of sand. A meteor plunges into our upper air with a relative velocity as high as 95km/s (an intrinsic speed of up to 65km/s added to Earth's own orbital motion of 30km/s); friction with air molecules causes it to glow red- then white-hot. It ionises the air it passes through, leaving a luminous trail that glows as a streak of light for up to a few seconds and can be detected by radar. A 'sonic boom' shock wave, due to air compressed before it, may be heard if the meteor is large enough to be called a fireball. If it explodes in flight it is a bolide. Meteors are much more likely to be seen after midnight than before.

Most meteors do not, surprisingly, start life in the asteroid belts. Some sporadic meteors may, but the regular meteor showers, named after the constellation in which their radiant lies – the Perseids, Leonids, etc. – were found by Schiaparelli in 1866 to occur when Earth crosses the orbit of a comet. Most meteors are caused by the low-density, fluffy silicates that are part of the make-up of a comet. Each time a comet's orbit brings it near the solar furnace it sheds some matter as its ices evaporate. Tiny, gritty, almost glassy fragments gradually spread around the comet's orbit, to be intercepted by Earth in its orbit.

This constant shedding of mass, over billions of years, seems to mean that we are rarely visited by a spectacular comet nowadays. The 'Great Comet' of 1843 spread its tail across many degrees of the sky like a searchlight beam, night after night; the comets of 1744 and 1861 had up to seven tails, spread fanwise. Most comets are named after their discoverer; no doubt the best-known is Halley's Comet, though Halley actually only worked out its orbit. Having discovered in 1704 that the orbits of comets are elongated ellipses, he then realised that comets recorded in 1456, 1531, 1607, and 1682 were one and the same, and that it returned roughly every 75 years. Its next return is due in 1986.

A comet consists of four main parts: the head or nucleus, up to a few tens of kilometres across, containing 'ices' of frozen methane, ammonia, carbon dioxide and water, mixed with perhaps 5 per cent of microscopic rocky particles; this is surrounded by the coma, which appears as a haze or glow around the nucleus, caused by ices vaporised by the Sun's heat. It can be anything from 10,000 to over a million km in diameter. The pressure of the solar wind forces out a tail – always away from the Sun, so that as the comet recedes from the Sun (for thousands or even a million years) its tail precedes it. The tail of the 1843 comet was 320 million km long.

There are actually two tails: a yellowish dust tail produced by solar radiation and usually straight, and a bluish plasma tail which curves with the magnetic lines of the solar wind. Finally there is the halo, an outer envelope surrounding the coma and visible only in ultra-violet light.

As it moves away from the Earth's vicinity the tail shrinks as all the volatile materials again solidify. A short-period comet, whose orbit is fairly circular, will return in a few years. The orbit of a long-period comet may be perturbed by a giant planet and become virtually parabolic, in which case it may never return; or its period may be shortened. Dutch astronomer Jan Oort suggested the vast spherical cloud of comets (the 'Oort cloud') in 1950. It surrounds the Solar System at a distance of 50,000AU and may contain 100 billion comets. At this distance the gravitational fields of other stars can affect or even capture them. And our Sun may lure away comets from other stars.

68

# Jupiter

- Mean diameter: 142,700km.
- Escape velocity: 60.22km/sec.
- Equatorial rotation period: 9hr 51min.
- Sidereal period: 11.86 years.
- Mean distance from Sun: 778,300,000km.
- Inclination of orbit: 1.3°.
- Inclination of equator to orbit: 3.08°.
- Mean density: 1.3.
- Surface temperature: −143°C.

*' It is one of the seven planets; its orbit is between Saturn and Mars
... and it is the biggest of all the planets.'*

THOMAS DYCHE and WILLIAM PARDON

## A New General English Dictionary (1744)

In the outer regions of the Solar System, which have been described frequently as intensely cold, where in the primordial nebula volatile compounds like ammonia, methane and water could condense and even freeze, there is a body that is second only to the Sun both in size and in emanations of energy; and yet, paradoxically, this body – the planet Jupiter – is being considered as a possible abode of life.

When it was formed by accretion, Jupiter was probably ten times larger than it is now; it shrank to about its present size by the time it was 10 million years old. If the building materials in the nebula had not run out when they did, and Jupiter had grown about 70 times more massive, then according to the mechanism by which bodies grow hotter as they grow larger (or as they shrink) it would have turned into a nuclear furnace and radiated as a minor star. The Sun would have become a double or binary star and the Earth and the rest of the planets would not be here at all. As it was, Jupiter had a core temperature of 50,000K, and it glowed, though only 1 per cent as bright as the Sun today.

Today, Jupiter's temperature (at the top of the cloud deck) is around 130K (−143°C), day or night, but there are 'hot spots' – holes in the cloud cover where the temperature is up to 260K. A 'black body' at the distance of Jupiter which absorbs all radiation falling on it would be at a temperature of 105K. The obvious implication is that Jupiter still radiates internal energy of its own. The core is probably still at some 30,000K – though there may be no core as such; Jupiter is very probably entirely liquid and gaseous with a density of only 1.3, compared with about 5 for the rocks of Earth and the other inner planets. We know that this could only be because Jupiter is composed purely of light elements – some 82 per cent hydrogen and 17 per cent helium in fact, with 1 per cent consisting of nitrogen, carbon

and a few other elements – in fact, the stuff that stars are made of.

The first telescopic observers after Galileo saw a flattened, cream-coloured disc crossed by a darker band or two. Larger instruments show that there are several reddish-brown belts separated by light 'zones', and odd streaks and spots are visible from time to time, both light and dark; and a special spot.

The British astronomer Robert Hooke first observed in 1664 the feature that has fascinated astronomers ever since – the Great Red Spot. At times it has vanished, only to reappear, and it has changed colour from pale pink through to deep brick-red. The Spot, a great oval, is so huge that three Earths could be lined up across it. Its position varies, it can move backward or forward in longitude by hundreds of degrees within a few years. This led early observers to believe that it was a solid object floating in a gaseous or liquid sea. It was even suggested that they were witnessing the birth of a new moon!

---

'Galileo wrote to Kepler wishing they could have a good laugh together at the stupidity of the ... professors of philosophy, who tried to conjure away Jupiter's moons, using logic-chopping arguments as though they were magic incantations.'
*A History of Western Philosophy* Bertrand Russell (1946)

---

Material within the Spot rotates anticlockwise about once every six days, at many hundreds of kilometres per hour. As for its colour, one theory suggests that it may be due to red phosphorus, produced by the action of ultra-violet light on phosphine gas brought up from below by currents around the Spot.

There has long been speculation about whether Jupiter has a solid core of iron and silicates, but many theorists today doubt this. Current theories put forward about the source of Jupiter's internal energy suggest that it must be

gravitational; the tiniest shrinkage – less than *1mm per year* – could account for today's observed temperatures.

Whether there is or is not a core, we do know that the interior is chiefly composed of hydrogen, at increasingly high pressures and temperatures with depth. Toward the centre it is transformed into a fluid but metallic state (the hydrogen is so condensed that it behaves like a metal). At the boundary is a transition zone, outside which the hydrogen is in a liquid but molecular state. In the upper few thousand kilometres it becomes a 'normal' gas, merging at the top with the cold atmosphere. In the atmosphere the lowest cloud-layers are believed to be of water droplets, then above those are ice crystals, ammonium hydrosulphide, ammonia 'cirrus', and finally a thin hydrocarbon 'smog'. At the top of the ammonia layer the pressure is down to about 1 atmosphere.

The ammonia cirrus clouds in the zones are white in colour, but the amazing variety of colours in the belts is still not fully explained. These were revealed in marvellous detail by the Voyager probes in 1979.

The colours could be caused by 'trace' impurities in otherwise white clouds. Different molecular forms (polymers) of sulphur can be yellow, brown or red – though no elemental sulphur has been detected. The widespread yellow colour could be caused by ammonium sulphide. Or ultra-violet radiation from the Sun may cause a reaction in some substance as it reaches the cloud tops. As it descends it returns to its colourless state, only to rise again in the zones in a continuous process of convection.

The orbital period of Jupiter is 11.86 years, and its orbit is inclined by 1°18′ to the ecliptic, while its axis is tilted by only 3°05′ to the perpendicular. Jupiter spins in a more nearly vertical position than any other planet except Mercury.

Its rotation is once every 9 hours 51 minutes – faster than any other planet,

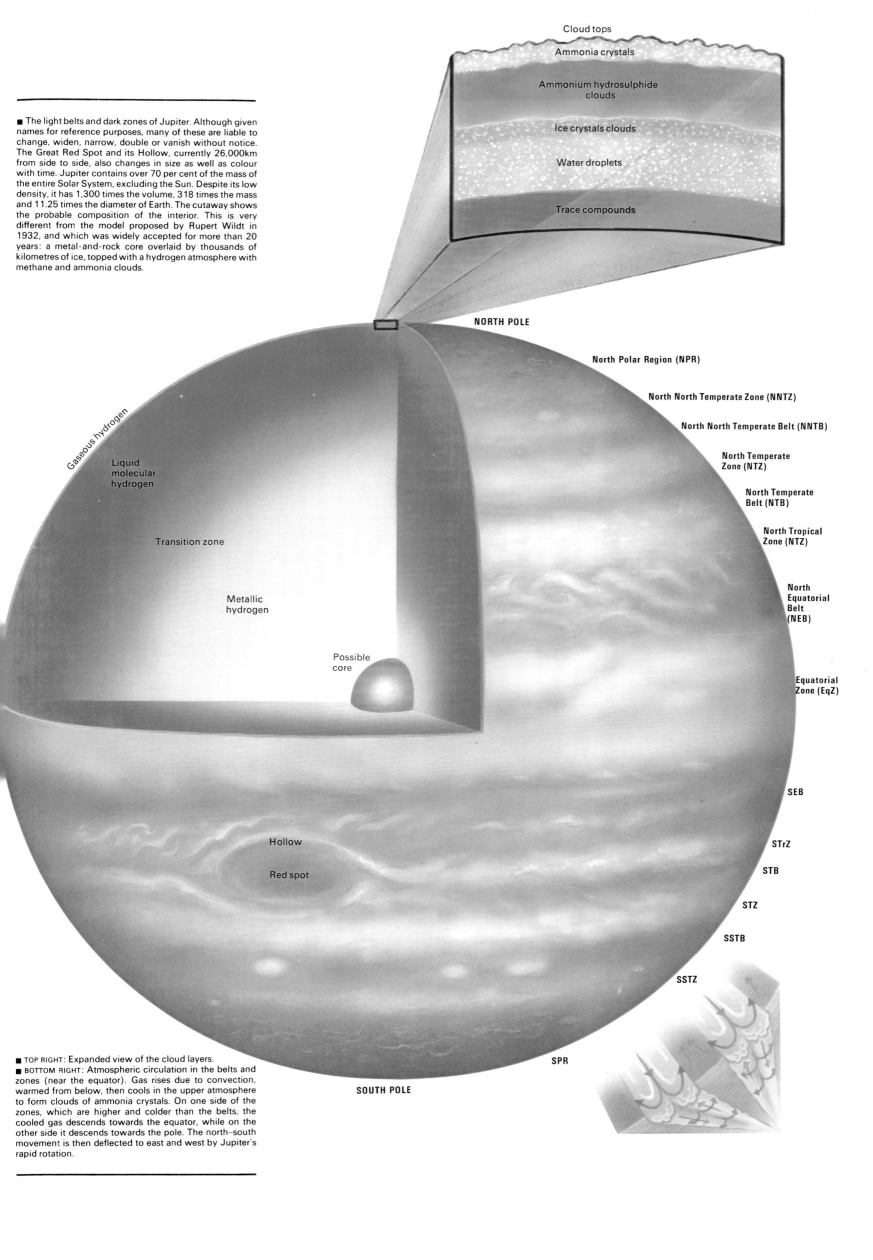

Cloud tops
Ammonia crystals
Ammonium hydrosulphide clouds
Ice crystals clouds
Water droplets
Trace compounds

■ The light belts and dark zones of Jupiter. Although given names for reference purposes, many of these are liable to change, widen, narrow, double or vanish without notice. The Great Red Spot and its Hollow, currently 26,000km from side to side, also changes in size as well as colour with time. Jupiter contains over 70 per cent of the mass of the entire Solar System, excluding the Sun. Despite its low density, it has 1,300 times the volume, 318 times the mass and 11.25 times the diameter of Earth. The cutaway shows the probable composition of the interior. This is very different from the model proposed by Rupert Wildt in 1932, and which was widely accepted for more than 20 years: a metal-and-rock core overlaid by thousands of kilometres of ice, topped with a hydrogen atmosphere with methane and ammonia clouds.

NORTH POLE

North Polar Region (NPR)

North North Temperate Zone (NNTZ)

North North Temperate Belt (NNTB)

North Temperate Zone (NTZ)

North Temperate Belt (NTB)

North Tropical Zone (NTZ)

North Equatorial Belt (NEB)

Equatorial Zone (EqZ)

SEB

STrZ

STB

STZ

SSTB

SSTZ

SPR

Gaseous hydrogen

Liquid molecular hydrogen

Transition zone

Metallic hydrogen

Possible core

Hollow

Red spot

SOUTH POLE

■ TOP RIGHT: Expanded view of the cloud layers.
■ BOTTOM RIGHT: Atmospheric circulation in the belts and zones (near the equator). Gas rises due to convection, warmed from below, then cools in the upper atmosphere to form clouds of ammonia crystals. On one side of the zones, which are higher and colder than the belts, the cooled gas descends towards the equator, while on the other side it descends towards the pole. The north–south movement is then deflected to east and west by Jupiter's rapid rotation.

which results in considerable flattening. Its equatorial diameter is 142,796km, its polar diameter 135,516km; but the equator rotates about 5 minutes faster than the temperate regions. Due to this rapid rotation there are powerful winds at the boundaries between zones and belts; there are also many jet streams, travelling at up to 150 metres per second. Some of the larger spots, which persist for years, may have 'roots' deep in Jupiter's fluid interior, what we see being the tops of columns. Lightning abounds on Jupiter; many 'superbolts' were photographed on the night side by Voyager 1. And it might be the action of this on methane and ammonia that causes orange and yellow organic polymers. The most provocative theory about the colours on Jupiter came in 1959 from Carl Sagan who, with his colleague Stanley Miller, passed sparks through a mixture of gases like those on Jupiter. The results were similar to those in Miller's original 'early Earth' experiment (page 30), and the experimenters concluded that "very large quantities of organic molecules must exist on Jupiter today".

In a more recent series of experiments, which attempted to reproduce the high pressures on Jupiter, two scientists claimed that the synthesis of organic molecules has reached an advanced state on Jupiter, and admitted the possibility that Earth-type microorganisms might be able to exist there. The idea is reinforced by studies of a living bacterium which is related to an organism that existed on Earth over 2 billion years ago, when the atmosphere was similar to Jupiter's.

But Sagan has gone even further and, with E. E. Salpeter, visualised enormous 'floating gas-bags', warming their interiors to keep aloft (like hot-air balloons) by feeding on the organic molecules that fall from the skies like manna from heaven. They might, he says, propel themselves with gas-jets.

No one, however, is seriously suggesting that any Jovians would be intelligent. Some scientists point out that the constant convective motions, from warm interior to cold upper atmosphere would quickly destroy any complex organic molecules that did form.

Radio waves – originally as interference or 'noise' – have been picked up from Jupiter from the 1950s onward.

## Radio Waves from Jupiter

There are two main types of radio waves. The long-wave radiation, the first to be discovered from Earth, came from the planet itself, while short-wave radiation was caused by electrons moving very rapidly through a magnetic field. The presence of such a field was again confirmed by the Pioneers and Voyagers. As might be expected, the field is extremely powerful; 10 times as strong as Earth's. Like Earth's, it is offset from the axis of rotation, by 11°, but it is *upside down* – our compasses would take us to the south pole instead of the north. The field is thought to be generated by electrical currents in the metallic hydrogen, in a similar way to those in Earth's rotating iron core. Aurorae were photographed near the poles by Voyager in huge arcs on the night side.

This great magnetic field results in a vast magnetosphere, similar to Earth's in basic structure but on a huge scale. There is an inner, doughnut-shaped belt containing trapped electrons and protons, extending outward about 10 Jupiter-radii. Beyond this is an outer belt, in which high-energy electrons are spread widely, but due to the rapid rotation it forms a flattened 'current sheet' extending over an area 10 times the diameter of the Sun, parallel to Jupiter's equatorial plane.

The Galilean moons (named for Galileo, their discoverer) orbit *inside* these belts and soak up radiation from them (especially Io), reducing the total radiation near the planet considerably.

Of at least as much interest as the planet itself during the Voyager fly-by are its satellites. By 1980, the number was 16. Some are probably no more than captured asteroids, and there may be many more to be discovered, but they are of no real significance. Apart from the Galileans, the satellites were numbered (not named), in order of discovery. So from the planet outward we have V, I (Io), II (Europa), III (Ganymede), IV (Callisto), VI, VII, X, XII, XI, VIII and IX, the last four orbiting in a retrograde direction and at high inclinations. The names given to the numbered moons are included in the tables on page 92.

In 1971 I wrote in the *Journal of the British Astronomical Association:* " . . . it is obvious from the variety of albedos, colours, densities and eccentricities of the satellites of the outer planets that some very strange sights could greet our eyes when eventually we do get there, and each world will have its own 'personality'." I had no idea how right I was! The pictures sent to Earth by Voyager 1 in March and by Voyager 2 in July, 1979, surpassed everyone's wildest expectations. As one scientist said, it was like seeing a new planetary system. Details of the moons examined, and interpretations (not necessarily the only ones) of the findings, starting at the outer Galilean, are as follows:

CALLISTO. Twice the size of our Moon, this proved to have the most familiar-looking surface, though it is the most intensely cratered body yet seen; very few craters are more than 150km across. Some have bright rims, perhaps of water-ice. It seems that virtually since it was formed Callisto has been geolog-

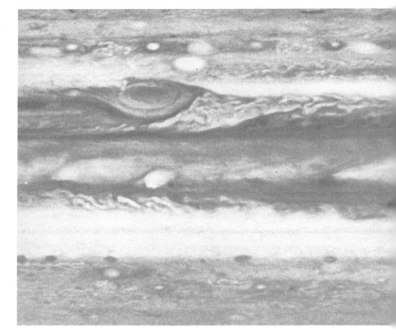

 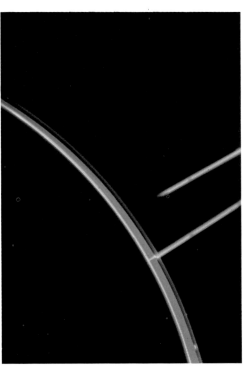

■ TOP: The north pole (left) and south pole (right) of Jupiter. Voyager did not fly over the poles, but these polar projections were reconstructed from several Voyager images. It is clear that near the poles the 'stretched-out' depressions and high-pressure areas break up, forming many minor whorls, spots and hurricanes more closely akin to those in Earth's atmosphere. (The black areas represent missing data.)

■ ABOVE, LEFT: The Great Red Spot, photographed by Voyager 1 from a distance of 4.3 million km. The smallest details here are about 100km across. To the south of the Spot is a large white spot. Such spots may persist for years, and could have 'roots' deep in Jupiter's fluid interior, what we see being the tops of columns. According to Dr. Garry

Hunt, the Red Spot is just the largest example of this uniquely Jovian phenomenon. If phosphine (PH₃) is brought up preferentially by the reddish spots they may be deeper-rooted than the white ones.

■ ABOVE, RIGHT: The rings of Jupiter, seen 'backlit' with the Sun eclipsed by Jupiter. The rings appear brightest in these conditions; a haze layer in the upper atmosphere scatters sunlight, but the image is misaligned.

■ BOTTOM LEFT: Jupiter, as photographed by Voyager 1 from a distance of 54 million km.

■ BELOW: A cylindrical projection of Jupiter, as if one could 'unroll' the planet like a map. This is how the great weather systems appeared early in July 1979.

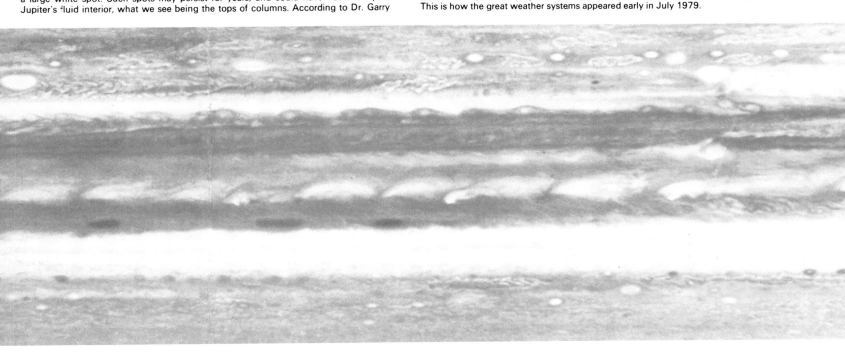

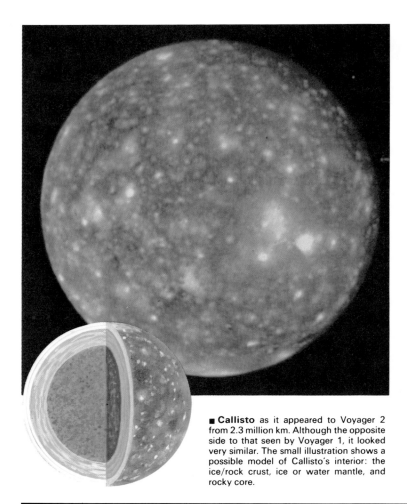

■ **Callisto** as it appeared to Voyager 2 from 2.3 million km. Although the opposite side to that seen by Voyager 1, it looked very similar. The small illustration shows a possible model of Callisto's interior: the ice/rock crust, ice or water mantle, and rocky core.

■ **Ganymede** from 2.6 million km, taken by Voyager 1. The smallest features are 50km across. The cutaway shows a similar composition to Callisto, but the ice crust thinner and the young, grooved terrain shows intricate fractures. The 2:1 orbital resonance with Io causes tidal heating effects.

ically inactive, and that it has a thin ice/rock crust over a water or ice mantle – a 'frozen ocean' in effect – and a rocky core. Ice flows have obliterated any large craters, and some craters, from their appearance, may have turned to slush before they re-froze.

Some craters are surrounded by concentric rings or 'annuli' – pressure patterns, or the 'ghost' remains of huge, old impact basins. Callisto has a relatively dark surface with light spots where, presumably, ice shows through. The maximum daytime temperature is −118°C; the minimum −193°C.

GANYMEDE, the largest moon, seems to have a somewhat similar internal

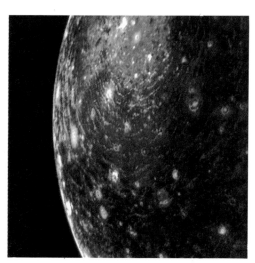

■ The old impact feature Valhalla on Callisto. The concentric rings extend 1,500km from a 600-km crater.

composition to Callisto, with a density of 1.6: half rock, half ice. But its surface is much more diverse, large areas of it covered by strange groove-like patterns – light bands of various widths. There are several dark areas, of which the largest is Regio Galileo, about 4,000km across. This bears traces of Callisto-type ghost basins, but the lighter, banded areas suggest much greater internal activity. There are parallel rows of mountains and valleys up to 1,000m high, sometimes between areas of darker, older surface, as if the crust had been stretched or pulled apart. Probably as the ice froze it expanded, pushing up whole sections of the crust into grooves and ridges.

There are some lightly cratered areas, and many of the larger craters have brilliantly white rays or haloes as if water or ice had been splashed from them. The oceans seem to have frozen at the time impact cratering stopped.

EUROPA, just a little smaller than the Moon, is covered by a frozen ocean, rather deeper than Earth's hydrosphere, over a large rocky interior. It is smoother, relatively, than a billiard ball, but looks as if it is covered in cracks. There are almost no impact craters; only three, about 20km across, are visible. Europa, then, would seem to have remained 'young' and active long after the bombardment of bodies in the Solar System ceased, and many meteorites would have splashed into a still

liquid ocean, which only froze much later.

The intersecting dark lines, which bear a strong resemblance to cracks in polar ice on Earth, were probably caused by the expansion of Europa as the ocean froze. The yellowish colour suggests that the ice is contaminated by pulverised rock fragments. However, the dark streaks are not cracks in the sense of being depressed below the surface; only 10 per cent darker than the ice they criss-cross, they are quite smooth as if painted on, suggesting that they became filled with material from below.

Most curious of all are the peculiar light ridges, all of which are curved into arcs, giving a regular, scalloped pattern. They stand a few hundred metres above the ice-plain, and have so far defied explanation, but they are unique to Europa.

Io. Of all the Galilean satellites, Io has proved the most spectacular, surprising, and exciting. In size it is similar to the Moon, and its density too is similar – 3.5. But not a single impact crater has been identified, despite Io's proximity to Jupiter's focus of gravity. There are many craters on Io, but all are volcanic – and a number of them are active!

This discovery was made by an engineer concerned with the navigation of Voyager 1. Linda Morabito was using her computer to enhance the image of a faint star when she noticed what she

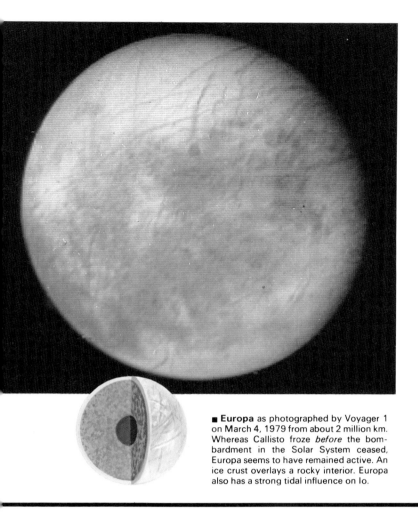

■ **Europa** as photographed by Voyager 1 on March 4, 1979 from about 2 million km. Whereas Callisto froze *before* the bombardment in the Solar System ceased, Europa seems to have remained active. An ice crust overlays a rocky interior. Europa also has a strong tidal influence on Io.

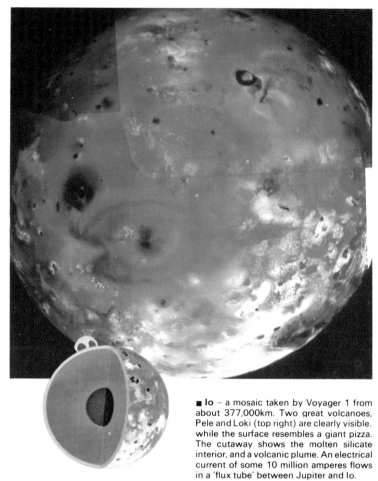

■ **Io** – a mosaic taken by Voyager 1 from about 377,000km. Two great volcanoes, Pele and Loki (top right) are clearly visible. while the surface resembles a giant pizza. The cutaway shows the molten silicate interior, and a volcanic plume. An electrical current of some 10 million amperes flows in a 'flux tube' between Jupiter and Io.

described as "an anomalous crescent" behind the limb of Io. At first she thought it must be one of the other satellites, then a new one, then a camera effect; each was ruled out in turn. Therefore it had to be connected with Io itself. It proved to be a plume from an active volcano, and its longitude and latitude identified it as originating from a heart-shaped feature named Pele – a volcanic caldera. It was the first erupting volcano ever observed on a body other than Earth. Eight or nine such volcanoes have now been photographed, their plumes of ejected material shooting from 70 to over 250km into space and falling back to the surface. And

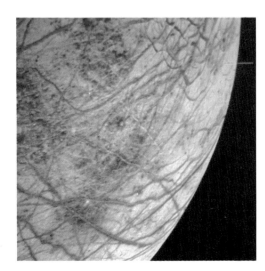

■ Voyager 2 took the first close-ups of Europa. Resolution here is 5km; the fracture patterns are clear.

what a surface! It is a variegated conglomeration of reds, yellows and whites, freckled with the black splotches of volcanoes.

Apart from the diffuse ejecta from explosive eruptions, fluid lava flows also play their part in a continuous resurfacing process. It is estimated that the surface we see today is not more than 100 million years old. A layer up to 1mm thick is spread over the entire surface each year; at that rate Io must have been turned completely inside out since its formation. The volcanoes, according to one main theory, are caused by sulphur dioxide ($SO_2$) boiling – at many degrees below zero – in contact with sulphur.

Only a thin crust of rock and sulphur, perhaps 300m thick, is skimmed on to a ball of molten lava; the boiling points of sulphur dioxide and sulphur differ so much that when liquid $SO_2$ mixes with sulphur it flashes into vapour, expanding greatly in volume below the crust and forcing its way to the surface in a rocket-like blast. Once there, it expands still more in the near-vacuum and forms an 'umbrella' of $SO_2$ gas and sulphur particles, creating a thin and variable atmosphere. If there is a fracture or fissure in the surface, molten sulphur may simply flow out.

Common 'flowers of sulphur' is pale yellow; at the temperature on the surface of Io ($-146°C$) almost white. When heated it turns orange, and it

retains that colour if cooled rapidly. This is known as an allotropic form. As more heat is added it becomes redder and also very fluid. So the orange and red flows visible on Io are caused by the runny forms of sulphur that froze, retaining their colours. There are 'lava lakes' on the surface, at 20–30°C. At still higher temperatures sulphur is deep red and viscous. This slow-running sulphur forms a 'flow front', causing cliffs or scarps and layered terrain. Where an outflow of liquid sulphur has undercut a plateau or mesa (one such plateau has an area of some 100,000 square kilometres), the layers are visible. There are also narrow valleys where the crust has subsided, as in terrestrial faults or graben.

In its final allotrope sulphur is brown-black, and this is the colour found on the floors of the calderas, which may be heated to some hundreds of degrees Celsius, although only 17°C was detected by Voyager's infra-red sensors. A crust of solidifying sulphur might account for this.

What is the energy-source for all this activity? Like the Moon and Earth, Io raises tides in Jupiter's atmosphere, and Jupiter exerts a massive pull on Io's interior. But unlike our Moon, slowly spiralling out of its orbit because of dissipation of its energy, Io *cannot* move further out because the orbits of the Galileans hold it firmly in place to be alternately squeezed and stretched by

the powerful Jovian tides. The heat generated by this constant deformation has melted and stirred Io's interior for billions of years. Water, carbon dioxide and methane were driven out of the rocks, leaving a sulphurous residue.

The plumes cool rapidly as they rise, and particles of sulphur and solid $SO_2$ fall back like snow within half an hour or so, creating big asymmetric rings around the vent. Often there is no cone as such. An estimated 100,000 tonnes of this material is erupted each *second* and returns; but several tonnes per second escape, some forming an ionosphere and some trailing behind Io like a 'wake'.

AMALTHEA (Jupiter V), the inner minor satellite which has been known since 1892, was shown by Voyager to have a very dark, red surface. It is 265km long by 150km wide, with its long axis always pointing toward Jupiter as it orbits every 12 hours; its red colour

may have something to do with the sulphur from Io. Inside Amalthea's orbit is the 14th satellite, Adastea, which was discovered in 1979 and loops around Jupiter at only 134,000km, faster than the planet revolves.

On 4 March 1979 Voyager 1 took an 11-minute exposure with its narrow-angle camera. It captured several star-trails – and the first image of Jupiter's rings, totally unexpected by current theories of planetary formation. Later photographs by both probes showed the rings clearly, especially when back-lit, with Voyager 2 in the planet's shadow. Unlike Saturn's broad ring-system, they are ribbon-like and only a few thousand kilometres wide, but *may* extend all the way to Jupiter's surface.

A follow-up to Voyager is planned by NASA: 'Galileo', a two-craft mission consisting of an orbiter and a probe to parachute into Jupiter's clouds.

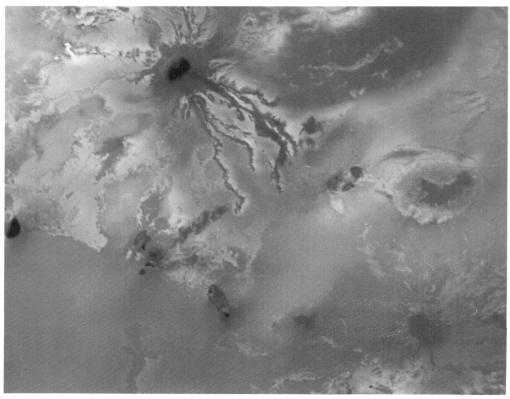

■ ABOVE: Recent volcanic activity is indicated by the dark red flows radiating from the dark centre. This picture of Io was taken from 130,000km; from top to bottom is about 1,000km.

■ BELOW: How a violent plume may be forced out, or molten sulphur may simply flow onto the surface of Io.

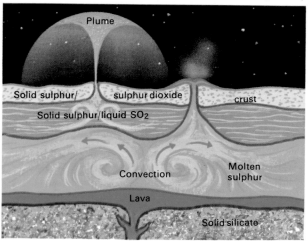

■ RIGHT: Dawn on Io. As the tiny Sun rises, a thin 'frost' is visible. This consists of sulphur dioxide, ejected during the night by volcanoes and frozen; it quickly vaporizes. A volcano splatters molten sulphur, while a distant plume rises kilometres into the sky. The thin crescent of Jupiter is extended by its atmosphere; its night side glows faintly in the light of several moons, but lightning 'superbolts' also illuminate the clouds, while great aurorae flicker at the poles. Jupiter remains fixed in Io's sky, but the Sun will shortly be eclipsed, to reappear and cause the planet to cycle through its phases to 'full' before sunset. A yellow glow will soon permeate the sky as energetic particles from Jupiter's magnetosphere ionise atoms of sulphur, oxygen and, especially, sodium, forming a 'plasma torus' and sodium cloud spread around Io's orbit. Being aligned with Jupiter's *magnetic* equator, these oscillate above and below Io's orbit.

# Saturn

- Mean diameter: 120,000km.
- Escape velocity: 36km/sec.
- Equatorial rotation period: 10hr 14min.
- Sidereal period: 29.46 years.
- Mean distance from Sun: 1,427,000,000km.
- Inclination of orbit: 2.49°.
- Inclination of equator to orbit: 26.73°.
- Mean density: 0.7.
- Surface temperature: −176°C.

*'We have learned more about the Saturn system in the past week than in the entire span of recorded history.'*
Scientist at the Jet Propulsion Laboratory, Pasadena (1980)

Saturn may no longer be unique in the Solar System, but to Earthbound observers it remains 'lord of the rings'. After the excitement of first seeing the Moon's craters, nothing can compare with one's first sight of this distant planet floating in the field of view encircled by its impossible halo. The ancients, to whom Saturn was the outermost planet, attributed slow-moving, sluggish qualities to it because of its tardy passage among the stars, and it was allocated the metal lead. This was ill-chosen, because Saturn, second largest planet in the Solar System after Jupiter, proves to be the least dense of any planet – only 0.7 the density of water, so it would float! By coincidence, using round figures Saturn is 9.5 times as far from the Sun as Earth, its diameter is just under 9.5 times that of Earth, and its mass is 95 times Earth's. If it were made of the same materials as Earth, the last figure would be 760 times.

Saturn's equatorial diameter is 120,000km, but its polar diameter is some 11 per cent less. This is partly due to the planet's rapid rotation – of 10 hours 14 minutes at the equator, but 26 minutes less at the poles – and also due to its low density. As with Jupiter, its atmospheric circulation varies at different latitudes. For Saturn is another basically gaseous planet; it is possible that it may have a small core of silicates and ices, and a smaller layer of metallic hydrogen than Jupiter. Most of its composition is probably of molecular hydrogen, either liquid or gas. Its cloud belts look similar, telescopically, to Jupiter's, but show less activity in the way of streaks and spots.

Galileo thought he saw a triple planet – a large one with a small one on each side – in 1610, then two years later the small ones had vanished. Later, he and other observers thought the planet had handles, and the term ansae (handles) is still used by astronomers. it was Huygens, in 1659, who explained all by

saying that Saturn is girdled by "a flat ring nowhere touching the planet". Cassini, in 1675, discovered the dark line which we now call Cassini's Division, though it was thought at the time to be a surface marking and it took William Herschel, over 100 years later, to prove that it was an actual gap or division between the bright Ring B and the outer Ring A.

'What a spectacle more worthy of admiration! On the one side (Mercury) we see thinking creatures among whom an Esquimaux or Hottentot would be a Newton, and on the other (Saturn) beings who would regard Newton with the same astonishment as they would a monkey.'
*Theorie des Himmels* Immanuel Kant (1755)

Herschel also measured the rings (the modern measurement is 274,000km from side to side: Ring A is 16,000km wide and Ring B 27,000km). He missed the faint inner Ring C or Crêpe Ring, as did several observers, and G. Bond in America and W. R. Dawes in England independently received credit for its discovery in 1850. But he did show that Saturn has an atmosphere, and discovered the two inner satellites, Mimas and Enceladus. Cassini had found Iapetus, Rhea, Dione and Tethys, between 1671 and 1684. For many years nine moons were known; today the number is uncertain, but at least 17 have so far been definitely identified.

The reason for the disappearance which mystified Galileo can be seen in the diagram right. Saturn's equator is inclined to the orbit by 26° 44′, and the rings are in the plane of the equator. Saturn makes a complete revolution of the Sun about every 29.5 years, so twice in each Saturnian year we see the rings 'wide open' (once for the northern surface, once for the southern); in between, they gradually close until we see them edge-on and they 'vanish' in a small telescope. The rings appear edge-on at alternating intervals of about 13.75 and 15.75 of our years; this oc-

curred in 1966 and 1979, and will next happen in 1995.

At first the rings were assumed to be a solid flat sheet or disc, but this was disproved in 1859 by the Scottish physicist James Clerk Maxwell. The mathematical basis for this was worked out some ten years earlier by a Frenchman, Edouard Roche. 'Roche's limit' is the distance between two bodies at which the smaller body will be disrupted by the tidal force exerted upon it.

An American astronomer, James Keeler, confirmed by spectroscopic observations in 1895 that Saturn's rings consist of a myriad of small particles, each orbiting like a miniature satellite. The Cassini Division is caused by the gravitational effects of the inner satellites, especially Mimas, which sweep particles from the division; but the Voyager 1 probe showed that the whole ring system is far more complex than anyone had suspected from Earth-based observations. That the particles consist of ice or ice-covered rock was shown by infra-red observations by G. Kuiper in 1949, explaining the rings' brilliance.

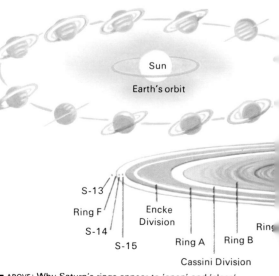

■ ABOVE: Why Saturn's rings appear to 'open' and 'close', as first explained by Huygens. The angle of Saturn's equator remains fixed relative to the stars as the planet slowly orbits the Sun.

■ ABOVE, RIGHT: The rings and cloud-markings of Saturn. There is a very faint ring, G, beyond F.

■ ABOVE: This ultra-violet image shows that Ring C and the Cassini Division are bluer than A and B. The structure of the rings can be seen, with a suggestion of the 'spokes', as well as the shadow of the rings on the planet.

■ RIGHT: One of the best Voyager 1 pictures of Saturn, taken from 18 million km. The rings have the relative thickness of tissue paper.

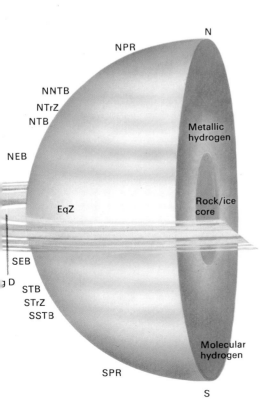

N

NPR

NNTB
NTrZ
NTB

Metallic
hydrogen

NEB

EqZ

Rock/ice
core

SEB

g D

STB
STrZ
SSTB

Molecular
hydrogen

SPR

S

■ ABOVE: A particularly beautiful (computer-enhanced) image, taken on 13 November, 1980 as Voyager 1 was leaving the Saturnian system, looking back at the crescent Saturn from 1.5 million km. The ring-shadow and light reflected onto the planet by the rings are visible, as is the outer, eccentric Ring F.

■ BELOW, LEFT: Detail in the belts of Saturn, including two brownish ovals some 10,000km across at about 40° and 60°N. A large red spot had appeared by the time Voyager 2 made its fly-by. ■ RIGHT: The moon S-15, visible here as a white dot, 800km outside Ring A and inside Ring F. 100km in diameter, this satellite appears to keep Ring A sharply defined.

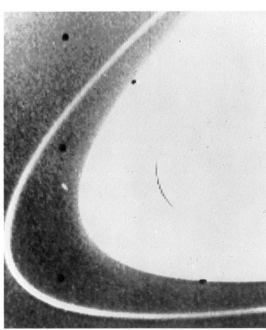

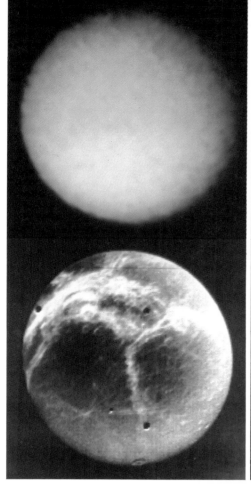

Similar explanations were advanced for the origin of the rings to those suggested for the asteroid belts: did an inner satellite approach unwisely close to Saturn's Roche limit and break up, spreading itself around the orbit? All the material in the rings would in fact make a very *minor* moon if gathered together. It is much more likely that tidal forces simply never permitted a proper satellite to aggregate from orbiting particles so close to the proto-Saturn in the early Solar System.

During the 1966 edge-on apparition, French observer A. Dollfus discovered the tiny tenth satellite close to the rings. Of the other satellites, about all that was known for certain before Voyager was that Titan is the largest satellite in the Solar System – larger even than Mercury – and has an atmosphere; the albedo of Iapetus varies greatly, one side being highly reflective and the other dark-coloured; the outermost moon, Phoebe, has a retrograde orbital motion and a high, 150° inclination; the five inner satellites, from Mimas to Rhea, are of very low density.

In November 1980 and August 1981, Voyagers 1 and 2 resolved the rings into *thousands* of structures from 10 to 50km wide. There are more rings *inside* Cassini's Division than we thought encircled Saturn! Ring F, detected by Pioneer 11, proved to be twisted and braided – "like a rope which has been twirled and is unwinding", said Dr. Garry Hunt. And it is flanked on either side by two new moons – S-13 and S-14 – 'herding' the particles into line. Then, when pictures of the two ansae were lined up, the main rings were also found to be eccentric.

But the biggest mystery was in the curious dark radial 'spokes' found in Ring A but not in B. It looks as though the material 'bunches up' as it orbits, and Saturn's magnetic field probably has some as yet unexplained effect.

The magnetic field itself was found by Pioneer and Voyager to be aligned with the rotation axis. Saturn's magnetosphere and radio emissions are similar to Jupiter's, though smaller; the bow shock is 26 Saturn-radii out.

The surface of Saturn is obscured by a hazy veil, but Voyager penetrated this on close approach. The belts and zones are less structured and well-defined than Jupiter's. There are brownish ovals and wave structures in some light bands, and several red spots, including one 5,000km long in the southern hemisphere. It is thought that sunlight reflected from the rings may affect the weather patterns, as may the shadow cast by the rings. Pioneer readings suggested that Saturn's heat-source was enormous, but Voyager data make it seem less powerful.

## Saturn's Moons

Several moons were seen for the first time as other than points of light. Five of the six listed here afforded few surprises, the sixth roused great interest. Starting at the innermost:

MIMAS is heavily cratered, and has one enormous crater with a central peak, on the Saturn-facing side. There are some large grooves among the many craters on the other side.

ENCELADUS is in part heavily cratered, but large areas are completely smooth and shiny, as though fresh ice had poured from its interior through cracks. It makes exactly two orbits for every one made by Dione, which may cause similar internal heating and melting effects to those in Jupiter's moon Io.

TETHYS is saturated with craters, but there are also huge fault valleys, one comparable with those on Mars – a definite sign of internal activity. Yet its density, like its companions', is only 1.2.

DIONE has bright, wispy streaks on the hemisphere opposite its direction of travel, which may be due to outgassing and freezing, with a very old impact basin at their centre. On the other hemisphere are many craters, some very large – and, again, faults.

RHEA is also very heavily cratered, and looks like the Moon or Mercury except that its craters are more irregular and the surface is much brighter. There are also large whitish streaks.

TITAN. This planet-sized body provided as much excitement as any other event of the fly-by. Visually, it is a featureless, hazy orange sphere with a bluish polar 'collar'. The first important discovery was of molecular nitrogen at the top of its dense atmosphere. The pressure may be up to twice that of Earth at the surface. Methane was confirmed, but perhaps only 1 per cent mixed with the nitrogen. The high haze is caused by ultra-violet solar radiation breaking down the methane, releasing hydrogen; a hydrocarbon residue settles to the surface. This of course opens up hopes for organic compounds and life – though reactions are incredibly slow at such low temperatures. Garry Hunt sums up by calling Titan "the Venus of the outer Solar System" and saying that it is "like having the Earth in cold storage . . ."

# Uranus and Neptune

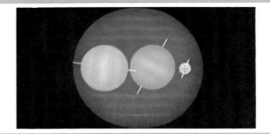

| | Uranus | Neptune |
|---|---|---|
| ■ Mean diameter: | 51,800km | 49,000km |
| ■ Rotation period: | 23 hours? | 22 hours? |
| ■ Sidereal period: | 84 years | 164.8 years |
| ■ Mean distance from Sun: | 2,869,600,000 | 4,498,000,000 |
| ■ Inclination of orbit: | 0.8° | 1.8° |
| ■ Inclination of equator to orbit: | 98° | 28.8° |
| ■ Mean density: | 1.2 | 1.6 |
| ■ Surface temperature: | −216°C | −228°C |

*'. . . We see that, arriving at the distant regions of Uranus and Neptune, the day star is reduced to dimensions which would hardly tempt us to transfer our household goods to these northern latitudes.'*

CAMILLE FLAMMARION Popular Astronomy 1894.

No spacecraft has yet visited Uranus. Voyager 2 will make a close approach in 1986, and if all goes well will continue on to reach Neptune in 1989. So for the moment, the temperatures of the Saturnian moons are the lowest we have encountered; for the first time we have encountered methane frozen into ice.

Imagine, then, the frigid conditions around Uranus, at a distance of 19.18AU compared with Saturn's 9.54AU – over twice as far from the Sun. This means that when William Herschel accidentally discovered Uranus in 1781 while mapping stars in the constellation Gemini (the first person to discover a new planet), he doubled the size of the known Solar System, and astonished the astronomical community. Herschel named it 'Georgium Sidus' (the Georgian Star) out of gratitude to King George III who, on hearing of the new discovery, had made Herschel a grant of £200 a year for life to enable him to continue his astronomical pursuits. Others felt that it should be named Herschel. It was Bode (of the Law) who suggested the name Uranus, after the mythological father of Saturn; and tradition finally prevailed, after several decades.

Herschel had found a strange world. He discovered four satellites in 1787 (Kuiper discovered the fifth in 1948. All are less than 2,000km in diameter), which moved up and down. Their orbits are inclined to their parent planet's by *more than a right angle* – 98°. It followed that Uranus' equator is inclined by the same angle, and that there must be an equatorial bulge. Having an axis which is tilted 8° below the horizontal means that the planet's normal anticlockwise rotation is technically retrograde – that is, when projected on to the orbital plane. As it rolls around its orbit, instead of spinning like a top, it shows observers on Earth first one pole, then a nearly vertical equator with a pole to each side, then the other pole. Uranus makes a revolution of the Sun in 84 years; its rotation was thought until

recently to be nearly 11 hours, but is now believed to be around 24 hours.

Methane was detected in 1932, although its absorption bands had appeared in spectra obtained in 1869. Molecular hydrogen was detected in 1952. The temperature at the surface of Uranus is 57K (−216°C) both theoretically and as measured, suggesting that there is no appreciable energy source.

It is not an easy object to observe, appearing as a pale bluish-green disc, on which faint equatorial belts and zones have been reported on occasion. However, observations in various wavelengths by the most sensitive instruments on Earth, including some made with a telescope carried to high altitude by balloon reveal no such markings, and this *may* be another case of observers seeing what they expect to see. The planet's colour is caused by the presence of methane, which strongly absorbs the red wavelengths of sunlight. The cloud layers may consist of solid particles of ammonia, which is expected to exist but not yet detected. Voyager should confirm this.

'In the quarter near Tauri the lowest of two is a curious either Nebulous star or perhaps a Comet.' Extract from Herschel's journal for Tuesday March 13, 1781 – his first sighting of Uranus.

Uranus has an equatorial diameter of 51,800km. According to one current model, Uranus has a rocky core up to 16,000km across, surrounded by an 8,000km-thick ice mantle, then molecular hydrogen – similar to a model for Jupiter proposed in the 1930s.

In the same year that he discovered the satellites, Herschel also detected polar flattening and suspected that he

saw rings, rather less in diameter than Saturn's Ring A. Although he recorded these several times, he was never quite convinced that they were not an optical illusion. No modern observations have shown the rings visually; yet oddly enough, Uranus is indeed a ringed planet.

On 10 March 1977, Uranus was due to occult (pass in front of) a star in the constellation of Libra. A number of professional astronomers, including an airborne American group, wanted to observe this. Photoelectric and other instruments were used in the hope of obtaining information on the atmosphere and size of Uranus as the star vanished and reappeared.

At several locations the star's light was seen to fluctuate a number of times both *before* and *after* the actual occultation. It was at first assumed that this was caused by several small unknown moons, but the symmetrical nature of the fluctuations was later analysed as being due to a system of five rings. They were given Greek letters from Alpha to Epsilon. The outermost and largest, Epsilon, was said to be 100km across and the others 10km, with 1,000km divisions between each.

The inner ring is 42,000km from the planet and the outer 51,000km. Later observations revealed four more rings within the group, making nine in all, with sharply defined edges. Epsilon seems to have a highly complex structure and is probably elliptical, varying from 20km wide at its closest to Uranus to 100km at its farthest. The rings are probably composed of dark, rocky (possibly chondritic) particles quite different from Saturn's icy hailstones. More recently it was shown that the rings may in fact be horseshoe-shaped. First Saturn, then Uranus, and now Jupiter have been shown to possess rings – it seems that the process of ring-making is not so rare as was once thought. Why is this? One theory by L. A. Wilson, a physicist, takes us back to the time when the outer, giant planets glowed with inner heat from contraction, and the T-Tauri wind swept through the Solar System.

The fact that Saturn's moons have progressively more mass the further they are from the planet suggests that first water-ice and then methane ice was allowed to freeze out as the planet cooled. Beyond Titan there was less material to form moons; and in the *inner* regions, only water-ice was allowed to condense to form the rings. But if there was still any gas left in the disc of material around the planet – as there should be – such particles would be slowed by 'drag forces', just as artificial satellites burn up due to air resistance.

When the Sun reached its T-Tauri stage, the wind blew away the remaining dust and gas, and any rings already formed would be safe. Uranus may once have had a much more impressive ring-system, lost some to drag forces, but retained some particles inside its Roche limit until the T-Tauri wind arrived.

However, S. Dermott and T. Gold of Cornell University and A. Sinclair of the Royal Greenwich Observatory, who first proposed that the rings of Uranus were crescent- then horseshoe-shaped, suggest that one or more satellites actually came within Uranus' Roche

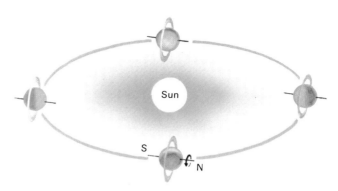

■ ABOVE: The strange seasons on Uranus, caused by its 98° axial tilt. On the left only the northern hemisphere receives sunlight, on the right only the southern; in between, more normal 'seasons' prevail.

■ BELOW: Uranus as it might appear from just beyond its third moon, Umbriel. An inner moon, Ariel, is visible against the edge-on rings, its crescent tilted at an odd-looking angle.

limit. But they formed several 'parent bodies' which remain in orbit; particles are constantly pulled from them by tidal forces, and kept in narrow orbits by the gravitational influence of the parent bodies, following horseshoe paths around the latters' Lagrangian points and causing the rings' eccentricity. This theory predicts that Neptune will also have similar rings, and that Phobos will break up to form a ring around Mars within a million years.

■ Neptune as seen from Triton. Its highly inclined orbit is taking the moon well above Neptune's southern hemisphere. An atmosphere (methane) is shown, though this is by no means certain and may be confirmed by Voyager 2's mission.

Earth may have had a ring, too, according to Dr. J. A. O'Keefe, a NASA scientist. He suggested that a great volcanic outburst on the Moon about 34 million years ago gave Earth a ring of glassy tektites (a type of meteorite found on both Earth and Moon). The ring would have lasted for a few million years, and its shadow on the Earth's surface would have caused dramatic climatic changes. Geologists know that a sudden climatic change occurred 34 million years ago, and call it the 'terminal Eocene event'. Many of the tektites would have fallen to the Earth's surface; the rest formed the rings,

eventually to spiral down due to gravitational forces. But by that time, several species of warm-water radiolaria had died out (Page 26). Tektites are found in the same sediments. Another theory suggests that a 10km asteroid hit Earth 65 million years ago, its dust blocking sunlight and leading to the extinction of the dinosaurs.

## NEPTUNE

During the early 19th century, several astronomers and mathematicians became intrigued by the fact that Uranus did not seem to obey Kepler's laws of planetary motion. John Adams, a noted

English astronomer, calculated between 1841 and 1845 that it was being affected by the gravity of yet another planet, farther out. In France, Urbain Le Verrier had arrived at the same conclusion. Unfortunately, both tried at first to use Bode's 'law', which forecast that the next planet should be 38.8AU from the Sun; as we saw, the law falls down at that point. Astronomers at Cambridge searched for the planet in 1846 – and actually saw it several times but failed to recognise it. As a result, the honours for discovering Neptune (on 23 September 1846) went to two young German astronomers, J. G. Galle and H. d'Arrest, using Le Verrier's calculations. It was the first of the planets to be accurately predicted. The actual distance is 30AU.

## Twin Planets

Physically, Uranus and Neptune are almost twins. For years Neptune was thought to be slightly larger, but a measurement in 1968 using the occultation method made Uranus 5 per cent larger, though 15 per cent less massive. The interiors and composition of both must be very similar. Neptune is rather bluer, and its surface temperature is only 45K; some internal energy may be produced by tidal interaction with its large, close satellite, Triton. This moon vies with Ganymede and Titan for the position of 'largest satellite', all being in the 5,000km class. Oddly, it orbits in a retrograde manner, at a very high inclination – 160°. (Neptune's equator is inclined by 28° 48'.) The orbit of the small outer moon, Nereid, is more elliptical than any other.

Neptune is the only planet about which nothing new or startling has been discovered in the last decade. Perhaps Voyager 2 will surprise us when it passes in 1989.

# Pluto

- Mean diameter: 2,600–3,600km.
- Rotation period: 6.4 days.
- Sidereal period: 248.5 years.
- Mean distance from Sun: 5,900,000,000km.
- Inclination of orbit: 17.2°.
- Inclination of equator to orbit: 57°.
- Mean density: 1.0?
- Surface temperature: −233°C?

*⁶There should exist . . . a large planet, sailing at 4,000 millions of miles from the sun in a revolution of 330 years . . . such is the last halting place in our planetary voyage; such is the last station of the sun's vast empire.⁹*

CAMILLE FLAMMARION Popular Astronomy 1894

Even after the discovery of Neptune, the motions of Uranus – and Neptune itself – still did not conform with the rules laid down by Kepler, and a search was started for a ninth planet. Lowell searched for 'Planet X' from 1905, but died in 1916, two years after publishing his calculations. W. H. Pickering (who discovered Saturn's moon Phoebe) and M. Humason continued the search photographically from Mount Wilson in 1919, without success.

Clyde Tombaugh of Lowell Observatory used a blink microscope, which causes any object that has moved, on otherwise identical photographs, to appear to 'jump'. After months of effort, he announced on 13 March 1930 that he had discovered Pluto, in Gemini. So although, like Neptune, it was found by mathematical prediction, photography was used for the first time to identify a new planet.

Or *was* it the expected planet? Pluto did not have the calculated mass – 6.7 times that of Earth was Lowell's prediction – that should have caused the perturbations of Uranus and Neptune which led to its discovery! Did the clue to the true character of Pluto lie perhaps in the Neptunian system? Certainly, something rather strange seems to have happened there. Neptune's large satellite Triton is only 354,000km from the planet, and according to one theorist, T. McCord, its orbit is rapidly decaying (spiralling in). And why is Nereid's orbit so comet-like?

In 1936 the British mathematician R. A. Lyttleton theorised that if Neptune once had *two* large satellites of about Triton's size in rather eccentric orbits, an encounter between them could have thrown one right out of Neptune's system, while the other – Triton – took up its present strange orbit. Kuiper has also put forward an alternative but similar hypothesis. No doubt the orbit of Nereid was perturbed also.

The orbital path about the Sun of the body which was ejected would be ex-

pected to be eccentric, too. This is certainly the case with Pluto, which in spite of being the outermost planet, can approach the Sun more closely than Neptune does; its next perihelion is in 1989. Its orbit is inclined to the ecliptic by 17°, too. This theory – and it is no more – is supported by the fact that Pluto is a solid body, not another gas giant, and is similar in its reflective characteristics to many of the satellites of the outer planets. But one might expect Pluto to approach Neptune more closely. (There is no chance of a collision because Pluto's orbit is so highly inclined; it loops more than a billion kilometres above the ecliptic plane, and never passes closer to Neptune than 18AU. Neptune's orbit is in a 3-to-2 orbital resonance with Pluto's.)

> 'It is unlikely that anything as large as the Earth is lurking beyond the present boundaries of the solar system . . . Gravitational theory looks unlikely to complete a hat-trick of planetary discoveries.'
> *New Science in the Solar System* Simon Mitten (1975)

Over the years since Pluto's discovery, its estimated size has varied between that of the Earth and that of the Moon. Its actual size has been difficult to determine, since it does not appear as a disc even in a large telescope, merely a point of light. The most recent estimates give 2,600–3,600km. Much depends upon its reflectivity; we could even be seeing a 'point reflection' of the Sun in a sea of frozen gas. Solid methane was in fact revealed by spectroscopic means in 1976; but its surface seems to be growing slowly darker.

The surface temperature is about 40K, varying from day- to night-side as Pluto rotates, in 6.4 days, according to measurements in 1955 from regular fluctuations in brightness, and by other methods since. A magnetic field therefore seems unlikely.

In 1978, James Christy of the US Naval Observatory noticed that an image of Pluto had a little 'bump' on

one side. On examining photographs taken in 1965 and 1970 it was found that these, too, showed the bump, in different positions to each side; these had been ignored as emulsion faults. The only logical interpretation was that it must be a large, close satellite, which Christy named Charon, after the boatman of mythology who ferried souls across the Styx. The first two letters of Pluto (who is god of the underworld) also commemorate Percival Lowell's initials.)

Charon orbits at about 17,000km every 6.4 days, making it synchronous with Pluto's rotation; it thus hangs over one spot on Pluto's surface. Pluto also rotates 'sideways', though not so steeply as Uranus; its equator is inclined by 57°, so Charon revolves at the same angle. The moon is about 1,100–1,500km in diameter to judge from an occultation made by Charon (but not by Pluto) of a faint star on 6 April 1980, as seen from South Africa. It is thus much closer to being a 'double planet' than Earth and Moon. Within the next few years Pluto and Charon will eclipse each other as seen from Earth, and we shall learn much more about both.

A theory was proposed in 1978 that Pluto and Charon are two pieces of what was once one body, pulled apart when a large (hypothetical) planet passed close to Neptune, which at that time had Pluto, Triton and Nereid as satellites. Having tidally disrupted the whole satellite system it then receded into a distant solar orbit, where it remains, unobserved. The theory was tested with computer 'models', which confirmed that the situation we see today could well result from such an encounter, and that Charon's synchronous orbit supports it. But then again, Pluto could be no more than a large agglomeration of primordial material which never *became* a moon, or just happened to be in a resonant solar orbit with Neptune instead.

Charon should not be confused with Chiron, much mooted in the press as a 'new' or 'tenth planet' when discovered by C. T. Kowal in November 1977. Chiron orbits between Saturn, whose orbit it crosses, and Uranus in a rather eccentric orbit somewhat reminiscent of Pluto's. If an icy body, as seems possible (it appears to resemble Phoebe) it is only 100km or so in diameter, but if composed of dark material, like Phobos, it could be as large as 800km across. It is certainly not large enough to be a planet, though.

However, A. T. Lawton of the British interplanetary Society has suggested that if Chiron proves to be carbonaceous, having condensed in the deeper reaches, there could even be a whole family of these 'dark asteroids', perturbing the orbits of outer satellites or being captured to form new ones.

Since Pluto did not seem to be able to account for its neighbours' quirks, a search for a genuine tenth planet began. Tombaugh himself conducted one, in the plane of the Solar System, and concluded that no planet of Neptune's size and mass could exist out to 100AU. J. L. Brady of the University of California has suggested that Halley's Comet, with two millennia of recorded observations, could aid in the search. The perturbation effects of all *known* bodies at each return were analysed by computer. The 'residuals' found to exist in the date of perihelion passage of the comet *could* be due to a trans-Plutonian planet.

According to Kowal, Galileo saw Neptune in 1612 and 1613 but took it for a star. From such ancient records it may be possible to find new clues to the position of the real Planet X, if it exists. Lawton had calculated a possible position from 'pre-discovery' observations of Neptune by the French astronomer J. J. Lalande; Galileo's observations, if accurate enough, are valuable because they come from a whole Neptunian year earlier. There is, of course, always the possibility of a large body orbiting high above (or below) the ecliptic, where little searching has yet been done.

A 'tenth planet' has also been reported several times *inside* Mercury's orbit. In the 1850s Le Verrier pointed out that the node of Mercury's orbit was precessing around the Sun by 43 seconds of arc each century. This he ascribed to perturbation by an intra-Mercurian planet, which he named 'Vulcan'. If such a planet were the size of our Moon it would be impossible to detect from Earth with a normal telescope, so close to the Sun, except during an eclipse or a transit. *If* it exists, it must be red-hot.

Le Verrier failed to observe such a transit, and Mercury's motion was explained in 1905 when Einstein's Special Theory of Relativity showed how the mass of a body depends on its speed relative to the speed of light. Despite reported observations during the eclipse of 1970, the existence of this worldlet is still dubious.

■ Looking back from several million kilometres beyond Pluto, Charon is about to be eclipsed by the planet; the distant, starlike Sun shines between them.

# Exploring the Future

■ First manned lighter-than-air flight: 1783.
■ First powered heavier-than-air flight: 1903.
■ First unmanned satellite: Sputnik 1, 1957.
■ First manned satellite: Vostok 1, 1961.
■ First unmanned soft-landing on Moon: Luna 9, 1965.
■ First manned landing on Moon: Apollo 11, 1969.
■ First manned space station: Salyut, 1971.
■ First unmanned landing on Mars: Viking, 1976.
■ First reusable spacecraft launch: Space Shuttle, 1981.
■ Next manned landing:?

*'As you hurtle from earth in a craft unlike any other ever constructed you will do so in the feat of American technology and American will.'*

PRESIDENT REAGAN to the US astronauts Crippen and Young in the Columbia Space Shuttle (1981).

It should be clear from the preceding pages that we owe an immense debt of gratitude to the scientists and engineers who design and control the space probes and their propulsion and imaging systems and analyse the stream of amazing data from them. But, without putting down the role of the robot craft, nothing can really replace the reactions of a 'man on the spot'. A robot explorer has to be fully automatic and pre-programmed; radio signals simply take too long for a human operator on Earth to control it remotely. A human explorer, though, may suddenly decide to investigate an interesting shape which catches his eye.

There are many practical reasons to continue our pioneering in space, apart from the urge to explore the unknown and to further scientific knowledge, so essential to Man's survival in the long run. Space offers huge advances in energy production, in industrial manufacturing processes (taking pollution away from the Earth, and offering 'free' heat, cold, vacuum, zero gravity, etc.), in education, communication, health, weather prediction and even climate control, and in discovering and shepherding resources on Earth and elsewhere in the Solar System.

Thousands of members of the L-5 Society in America are urging the construction of giant space colonies located on the Moon's orbit, equidistant from Earth and Moon, where there is perpetual sunlight. As well as building more colonies, their inhabitants would construct satellite solar power stations (SSPS) to provide electrical power derived from sunlight and beamed to Earth. To become cost-effective, material for later colonies would come not from Earth but from the Moon, delivered by an inexpensive magnetic catapult called a mass-driver.

Farther out in space, the asteroids are ripe for exploitation. As we have seen, many of these are composed of almost pure nickel-iron; there must be enough to supply Earth's needs for decades, at least. Many other materials will also be available: carbon, ice, hydrogen and methane may all be shipped in from the outer Solar System one day, when it becomes economically feasible to do so. New forms of propulsion, such as the electric or ion drive, will come into use, perhaps combined with solar panels to provide electrical power. (Unfortunately, NASA's Solar Electrical Propulsion System (SEPS) has been cancelled for the present, along with a Halley's Comet probe, the Solar Polar Mission and other projects. We must await the 1990s for a Mars Sample Return mission, a Dual Orbiter to Saturn and Titan, an Asteroid Multiple Rendezvous and a Mercury Low Altitude Orbiter and Lander.)

The pressure of sunlight can be used alone to propel a solar sail – basically an inexpensive sheet of thin metallised plastic film. But we do not need to wait for nuclear or other exotic forms of propulsion. For instance, Dr. R. C. (Bob) Parkinson of the BIS has proposed a conventionally fuelled five-man expe-

'Within a few thousand years of entering the stage of industrial development, any intelligent species should be found occupying an artificial biosphere around its parent star.'
Freeman J. Dyson (1966)

dition to Mars which could, technically, be launched in 1995, using Space Shuttle technology and hardware, together with the Space lab being built by the European Space Agency (ESA) and NASA's Heavy Boost Stages.

Beyond that, the technology certainly exists to build bases capable of supporting humans for extended periods, not just in Earth-orbit or on the Moon but in the alien environments of Mars or the moons of Jupiter or Saturn. Astronomers would benefit especially by being able to do their observing, in whatever wavelength, well away from the distortions and interferences on Earth. There is no problem in extracting oxygen and other necessary materials from lunar rock, and plants grown hydroponically (without soil) would supplement this as well as providing food.

Taking the concept of manned bases on other planets still further, there is the staggering possibility of 'planetary engineering' or terraforming – a term coined in 1942 by science fiction writer Jack Williamson. The idea is simply to make other worlds habitable by humans. An early suggestion, in 1961, by Carl Sagan was to 'seed' the atmosphere of Venus with blue-green algae, converting the carbon dioxide into oxygen and at the same time reducing the pressure and temperature (by eliminating the greenhouse effect). The upper clouds would condense and rain would fall, forming oceans.

A more recent alternative, now that we know how hostile Venus really is, is to ferry in ice asteroids 15km or so in diameter, put them into orbit around Venus and aim them, using rocket jets, at a specific spot on the surface. Each crashes at nearly 100km/s, at such an angle that Venus' rotation is increased until a 24-hour day is approached, while at the same time water is provided as the ice melts. *Then* the atmosphere is seeded with blue-green algae.

The same could even be done with the Moon; once given a breathable atmosphere by baking oxgen out of the rocks with giant parabolic mirrors, it would remain for thousands of years, even if not replenished. The time factor for the operation is remarkably short. Mercury would need to be shielded from the Sun by a 'parasol' of rocky particles put up by mass-driver, or by a man-made ring. Mars would need to be warmed up, perhaps by reflecting sunlight on to the poles with huge, thin metal-foil mirrors, increasing the energy-flow at the poles by 20 per cent. Or we could spread dark material from its carbonaceous moons on them with a mass-driver. Rich not only in carbon but in oxygen, nitrogen and hydrogen, this is excellent raw material for fertiliser. Once the atmosphere was thickened, the greenhouse effect and carefully chosen plant life should do the rest.

It all sounds like pure science fiction and, of course, at present it is. I am not advocating that we start the job today. But the above examples illustrate how sweeping is the imagination of Man; and what Man can imagine, he can accomplish – if the need is great enough.

One day, the Sun will finally have consumed its supply of hydrogen fuel, and will start to burn helium into carbon. When that happens – about 5 billion years hence – the Sun will expand into a 'red giant', glowing only red-hot but from a much greater surface area and so giving out more heat.

Not only the Earth but eventually all the planets will be roasted, then enve-loped as the sun swells. It is fascinating to speculate that Titan might, for a time at least, act as a 'lifeboat' for humanity as it thaws from its deep-freeze. But long before that interstellar travel will surely have been achieved. In the absence of a faster-than-light drive – which according to Einstein's Theory of Relativity is impossible – perhaps mined-out asteroids will be equipped with some form of propulsion and used as giant 'arks' to allow Man to spread throughout the Galaxy.

■ BELOW: The Sun casts long shadows as two members of the Mars expedition make an early morning start to explore the local hills in their Mars Rover. The landing module (designed by Bob Parkinson) is on the plain at left.

■ OVERLEAF: far in the future, Mars has been terraformed and given an atmosphere and climate suitable for Earth-people. Valles Marineris has become a real 'canal', with water and vegetation; on the night side, the lights of cities twinkle. We view the planet from a base on Phobos. At right is one of the mass-drivers which launched dark carbonaceous material to darken the poles.

# Glossary

**A**

**Absolute zero:** the lowest possible temperature, at which all heat motion of atoms ceases; 0 on the Kelvin scale, −273° Celsius (Centigrade).

**Absorption spectrum:** the dark lines, appearing against a bright background, caused when atoms between a source of light and the observer absorb characteristic wavelengths from the continuous **spectrum** of that light-source.

**Aeon:** one billion (1,000 million) years. (In USA, spelt 'eon'.)

**Aerolite:** a stony **meteorite**.

**Aeropause:** the transition zone between the **atmosphere** and space.

**Airglow:** faint luminosity in the atmosphere of Earth or other planet, caused by the excitation of air molecules by solar radiation.

**Albedo:** the reflectivity of a planet or other non-luminous body. A perfectly white object would have an albedo of 100 per cent.

**Altitude:** the angle of a celestial body above the **horizon**, from 0° at the horizon to 90° at the **zenith**.

**Ångström:** unit of measurement for the wavelength of **electromagnetic** energy, including light. 1Å (about the diameter of a hydrogen atom) is equal to one hundred-millionth of a centimetre; visible light ranges from about 7,500Å (red) to 3,900Å (violet).

**Angular diameter:** the angle subtended by an object at a given distance (eg. a planet as seen from its moon). It varies inversely with distance; eg. at twice the distance the angular diameter is halved.

**Anorthosite:** type of rock, of which much of the Moon's **crust** is composed (especially in the highlands). It is light in colour and rich in silica.

**Aphelion:** the farthest point from the Sun in the orbit of a planet or other body.

**Apogee:** the farthest position of the Moon or other body (including an artificial satellite) in its orbit around the Earth. It has been suggested that this term be used for any planetary body, in place of 'apoapsis' for Mars, 'apolune' for the Moon, and so on. (See **perigee**.)

**Apparition:** the period during which a planet is seen most favourably from Earth.

**Arc, degree of:** one hundred and sixtieth part of a complete circle (360°), usually as applied to the **celestial sphere**. Each degree is divided into sixty minutes (60′), and each minute into sixty seconds (60″).

**Asteroids:** several thousand minor planets (also known as planetoids) which move in orbits between those of Mars and Jupiter.

**Astronomical Unit (AU):** unit of distance based on the mean distance between the centres of Earth and Sun: 149.6 million kilometres.

**Atmosphere:** the shell of gas that surrounds some celestial bodies. It is normally densest near to the surface, and thins out gradually into interplanetary space.

**Atom:** the smallest unit of a chemical **element** able to retain its own individual character. It consists of a **nucleus** surrounded by orbiting **electrons**.

**Aurora:** a phenomenon of the upper atmosphere (on Earth, 80–300km), caused by the **ionisation** of the atoms in the atmosphere at this height by energetic particles from the Sun, trapped in the Van Allen radiation belts. Jupiter, in particular, also has vast auroral displays.

**Axis:** the imaginary line through the centre of a body, about which it rotates. At the surface of the body, at each end of the axis, is a pole (north and south).

**Azimuth:** the horizontal bearing of a celestial body, reckoning east or west from the north point of the observer's **horizon**. It constantly changes due to Earth's rotation.

**B**

**Barycentre:** the **centre of gravity** of the Earth-Moon system, around which both bodies revolve. This lies within the Earth, since the latter is 81 times more massive than the Moon.

**Basalt:** an igneous rock, very common on Earth, Moon and the **terrestrial planets:** low in silica content and often found in **lava** flows.

**Basin:** large circular depression on the **terrestrial planets** (and some of their satellites), presumably caused by an impact and flooded with basaltic **lava**.

**Big Bang:** the theory that the universe originated in a vast 'explosion', 12–20 billion years ago.

**Billion:** 1000 million. **Trillion:** one million million.

**Binary star:** a system of two stars orbiting about a common **centre of gravity**.

**Black body:** a body that absorbs all radiation falling upon it, reflecting none. (See **temperature**.)

**Black dwarf:** a small, star-like object whose temperature is too low for **nuclear** reactions to take place.

**Bolide:** a brilliant (and often an exploding) **meteor**.

**C**

**Carbon cycle:** a series of **nuclear** reactions within stars, using carbon and nitrogen as catalysts, in which hydrogen is converted into helium and energy released.

**Carbonaceous chondrite:** type of **meteorite**, rich in carbon and **volatiles**; may be the earliest material formed in the Solar System.

**Celestial sphere:** an imaginary sphere surrounding Earth, on which all celestial bodies appear to move. From the Earth, the **celestial equator** and **celestial poles** are projected onto the sphere.

**Centre of gravity** (or of **mass**): the point within any body or system at which all of its mass could theoretically be concentrated without affecting the balance or motion of the system.

**Coma:** the fuzzy patch round the **nucleus** of a **comet**.

**Comet:** a body, composed mainly of frozen gases, in a highly elliptical orbit round the Sun. It produces a tail when close to the Sun.

**Conduction:** a process by which heat energy is transferred from a hot region to a cooler one, as rapidly moving **molecules** in the hot matter pass on their motion to nearby molecules. (See also **convection** and **radiation**.)

**Conjunction:** the period when Earth and a planet are in line with the Sun; also used for the apparent close approach of one planet to another or to a star, as seen from Earth (a 'line of sight' effect).

**Convection:** the transference of heat by the motion of masses of material—eg. warmed air rising. Usually a continuous process of heating/rising, cooling/sinking.

**Core:** the deepest inner region of a planet or star.

**Corona:** the outer, tenuous **atmosphere** of the Sun.

**Cosmic rays:** high-energy atomic particles (chiefly **protons**) which enter Earth's atmosphere from outer space. On striking air molecules the primary rays split into secondaries, many of which reach the ground. The rays may originate in **supernovae**.

**Crust:** the outer, solid layer of a planet or satellite.

**Cusp:** one of the two pointed 'horns' of a crescent **phase**.

**D**

**Day:** the time taken for the Earth to **rotate** once on its **axis**; normally taken as 24 hours (but see **sidereal day**).

**Density:** the mass of a substance per unit volume. Water is given the value of 1.00.

**Dichotomy:** the exact moment of half-phase of the Moon, Venus or Mercury.

**Differentiation:** a process by which a mixture of chemicals is separated into different zones, as in the **crust** and **mantle** of a planet.

**Diurnal motion:** the apparent daily movement of celestial objects from east to west (actually due to Earth's **rotation** from west to east).

**Droppler effect:** the shift in the wavelength of light (also sound) as an object recedes (red shift) or approaches (blue shift), as seen by an observer. As the body approaches the apparent wavelength is shortened, and vice versa as it recedes.

**E**

**Eccentricity:** measure of the degree of ellipticity in the **orbit** of a body.

**Eclipse:** the passage of a celestial body into the shadow of another; eg. a **lunar eclipse** (Moon passes into Earth's shadow) or **solar eclipse** (Moon's shadow falls on Earth). Also used when any body is obscured by another.

**Ecliptic:** the projection of Earth's orbit onto the **celestial sphere**, inclined to the celestial equator by 23° 27′; or the apparent annual path of the Sun across the sky against the stars of the **Zodiac**.

**Ejecta blanket:** a layer of material thrown onto a planet's surface during the formation of a crater.

**Electromagnetic spectrum:** the full range of **radiation**, from gamma-rays, through visible light to radio waves.

**Electron:** a negatively charged particle orbiting the **nucleus** of an **atom**.

**Element:** a substance that has a specific number of **protons** in the nucleus of each of its atoms, and which cannot be split chemically into a simpler substance. 92 elements exist in nature, of which hydrogen (one proton) is the lightest, uranium the heaviest.

**Ellipse:** a conic section giving a closed, oval curve; the shape of the **orbit** of one body around another.

**Elongation:** the apparent angular distance of a planet from the Sun, or of a satellite from its planet.

**Equinoxes:** the dates (twice each year—21 March and 22 September) at which the Sun crosses the **celestial equator**, and night and day are of equal length; the points at which the **ecliptic** cuts the **celestial equator**.

**Escape velocity:** the minimum speed at which an object must be impelled in order to escape from the **gravitational** field of a body.

**Excitation:** the raising to higher energy levels of the **electrons** in an atom.

**F**

**First Point of Aries:** the vernal (spring) **equinox**. When the Sun crosses the **node** (where **celestial equator** and **ecliptic** cross) it is actually 'in' the stars of Pisces, not Aries, due to **precession**.

**Focus:** one of two points within an **elliptical orbit**; one is occupied by the Sun or other parent body, the other normally empty. Also, the point at which converging light rays meet, as in a telescope.

**Free fall:** the motion of an unpowered body under the influence of a **gravitational** field. It applies whether the object is actually falling, travelling between bodies in space, or in **orbit** about a body, providing no acceleration is applied. Any person or object within a vehicle in free fall is 'weightless', or experiencing zero gravity (zero-g).

**Frequency:** the number of oscillations per second in an **electromagnetic** wave of a given **wavelength**.

**G**

**g:** the symbol for the acceleration (increase in velocity), due to gravity, of a freely moving body at the surface of Earth—about 981 centimetres per second per second.

**Galaxy:** a system of millions of **stars**, plus gas and dust (and the possible planetary families of those stars). There are about 100 billion stars in our own **Milky Way** galaxy, and about a billion galaxies can be detected by our best telescopes.

**Gamma-rays:** **radiation** of very short wavelength—about a millionth that of visible light.

**Geocentric:** Earth-centred; measured with respect to the centre of Earth.

**Geodesy:** the science concerned with the shape, size, mass, etc. of Earth.

**Geophysics:** the science dealing with the physics of the Earth and its environment, from its interior to the outer **magnetosphere**.

**Gibbous:** **phase** of the Moon or a planet between half and full.

**Gravitation (gravity):** the force by which all masses in the universe attract each other.

**Greenhouse effect:** the trapping of heat within a planetary atmosphere due to the absorption of **infra-red** radiation.

**Greenwich Mean Time (GMT):** the time taken as standard worldwide; the time at Greenwich, England (longitude 0°). Also known as Universal Time (UT).

**H**

**Half-life:** the time taken for half of the **atoms** in a given sample of a **radioactive element** to decay (ie. change into a different element).

**Heliocentric:** centred on the Sun.

**Hertzsprung-Russell (H-R) Diagram:** a graph on which the spectral type (ie. colour, due to **temperature**) of stars is plotted against **luminosity**. There is a **Main Sequence** of stars from upper left (very hot, bluish stars) to lower right (cool, red stars), with branches for **giants** and **white dwarfs**.

**High-luminosity phase:** short-lived stage in a star's evolution, during which it is extremely bright, before it joins the **Main Sequence**.

**Horizon:** the line at which surface and sky appear to meet—normally a circle if no obstructions interfere. As it depends upon the curvature of the planet, the distance to the horizon varies; eg. on a flat plain on Earth it is about 5km, on the Moon 2.5km.

**Hyperbola:** an open conic section; the curved path of an object falling freely at a speed greater than **escape velocity**.

**I**

**Inferior planet:** one whose orbit is between the Earth and the Sun—ie. Mercury and Venus.

**Infra-red: radiation** longer than 7,500Å, up to short-wave radio wavelengths. It is invisible to human eyes, but may be felt as radiant heat.

**Interstellar:** between the stars of the **Milky Way** galaxy. May be applied to vehicles, bodies, dust, gas, molecules, atoms, etc.

**Ion:** an **atom** or **molecule** which has a positive or negative electrical charge because it has lost or gained **electrons**.

**Ionisation:** the process of removing **electrons** from an **atom** or **molecule**.

**Ionosphere:** region of the **atmosphere**, above 80km (on Earth), in which the atoms are ionised by particles in the **solar wind**, creating layers which reflect radio waves.

**Isotope:** forms of the same **element** in which the number of **neutrons** in the **nucleus** differ.

**J**

**Jovian planet:** the giant planets, which resemble Jupiter in composition.

K

**Kepler's Laws of Planetary Motion:** (1) All the planets move in **elliptical orbits,** with the Sun at one **focus,** the other being empty. (2) An imaginary line joining the centre of a planet or other orbiting body to the centre of the Sun sweeps out equal areas in equal times. (3) The square of the **sidereal period** of a planet is proportional to the cube of its mean distance from the Sun.
**Kinetic energy:** the energy a body has due to its motion.
**Kirkwood Gaps:** unoccupied zones in the **asteroid** belts. The orbit of any body entering a gap will be **perturbed** by the **gravitational** influence of Jupiter. First noted by US mathematician Daniel Kirkwood.

L

**Lagrangian point:** one of five points within any system consisting of one large and one smaller body in orbit, at which an object would remain in a fixed position with relation to the other two. The L-4 and L-5 points in the Earth-Moon system are much mooted for space colonies.
**Lava:** molten rock ejected from the interior of a planet onto its surface.
**Light-year** the distance travelled by light in one year, at 299,792.458km/sec; 9.4607 trillion kilometres.
**Limb:** the edge of the visible disc of any celestial body.
**Lithosphere:** the solid rocky layer 'floating' on the molten interior of a planet.
**Luminosity:** the total energy radiated by an object or light-source.

M

**Magnetosphere:** the magnetic field of Earth or other planet, including the tenuous gas (neutral and ionised) enclosed by it.
**Magnitude:** the measure of the apparent brightness of a star or other body. As it depends upon distance, it is not a true guide to **luminosity.**
**Main Sequence:** see **H–R Diagram.**
**Mantle:** the region of the interior of a planet which is outside the **core** and below the **crust.**
**Mare:** (plural **maria**) a dark, flat plain on the Moon or some planets.
**Mascon:** a concentration of mass found under several lunar **maria,** causing **gravitational** irregularities.
**Mass:** the amount of matter contained in a body (regardless of gravity, so not the same as **weight**).
**Mean Sun:** an imaginary Sun, moving at a constant 1° per day around the **celestial equator.**
**Meridian:** a circle on the **celestial sphere** passing through the **zenith** and the north and south points on the **horizon,** as seen by any one observer. Also, a north-south line on any celestial body.
**Meteor:** a 'shooting star'; the streak of light produced by a **meteoroid** burning up in the atmosphere.
**Meteorite:** a **meteoroid** large enough to reach Earth's surface after passing through the atmosphere.
**Meteoroid:** a particle in space, which produces a **meteor** or **meteorite** after being attracted by Earth's gravity into the atmosphere.
**Milky Way:** the diffuse band of light across the night sky; the name of the **galaxy** of which our Sun is a member.
**Molecule:** a group of two or more linked **atoms.**

N

**Nadir:** point on the **celestial sphere** directly delow the observer— the opposite to **zenith.**
**Nebula:** a cloud of gas and dust in space; if close to a star, it may become luminous. New stars are formed from nebular material.
**Neutrino:** a fundamental particle produced by some **nuclear** reactions in the heart of stars.
**Neutron:** an electrically neutral particle, whose mass is equal to that of a **proton,** in atomic **nuclei.**
**Node:** one of two points on the **orbit** of a celestial body where the orbit is crossed by the plane of the **ecliptic.**
**Nova:** a star that suddenly increases greatly in brightness, then returns to normal after days, weeks or months. The outer layers of the star are blown off, usually to be transferred to a **binary** companion.
**Nuclear fusion:** the process (by which stars release energy) in which two or more atomic nuclei fuse together to form a heavier **nucleus.**
**Nucleus:** the central 'hub' of a spiral **galaxy;** the bright central part of a comet; the **protons** and **neutrons** at the centre of an atom.

O

**Obliquity:** (of the ecliptic) the angle between the **ecliptic** and the **celestial equator**—23° 26′ 54″; the angle of any planet's **axis** to its **orbit.**
**Occultation:** the obscuring of one celestial body by another; when this is total, an **eclipse** occurs.
**Opposition:** the position of a **superior planet** in the sky (at midnight) when it is directly opposite the position of the Sun. The three bodies are thus in line.
**Orbit:** the closed path of any natural or artificial object around a larger body.

P

**Parabola** an open conic section; the curved path of an object moving at **escape velocity.** (Easily **perturbed** into an **eclipse** or **hyperbola.**
**Parsec;** a measurement of distance in space, equal to 3.26 **light-years.**
**Perigee:** the closest position of the Moon or other orbiting body to the Earth (see **apogee**).
**Perihelion:** the closest point to the Sun in the orbit of a planet or other body (see **aphelion**).
**Perturbation:** disturbance in the **orbit** of one body caused by the **gravitational** pull of one or more others.
**Phase:** the amount of the illuminated surface of the Moon or a planet, as seen from the earth or any point in space.
**Photon:** a unit of **electromagnetic radiation,** with a different amount of energy for each **wavelength.**
**Photosphere:** the Sun's visible, glowing surface.
**Planet:** a relatively large, non-luminous solid or gaseous body moving in an **orbit** around a **star.**
**Planetesimal:** small bodies (from millimetres to a few kilometres in diameter) which accumulate to form **planets.**
**Plasma:** gas at very high temperature consisting entirely of **ions** plus some neutral particles. It is a good conductor of electricity.
**Plate tectonics:** the motions in Earth's **crust** of large plates, causing continental drift, sea-floor spreading and the formation of mountain ranges.
**Precession:** slow conical motion or 'wobble' in the rotation **axis** of a celestial body.
**Prominence:** a mass of luminous gas (mainly hydrogen) extending from the surface of the Sun into the **corona.**
**Proton:** a positively charged fundamental particle in the **nucleus** of an **atom** (see **neutron**).
**Protoplanet:** a planet in the early stages of its formation.
**Protostar:** a large mass of gas, about to become a **star,** before **nuclear** reactions commence.

Q

**Quadrature:** half-phase; the position of the Moon or a planet when it is at a right-angle to the Sun, as seen from Earth.
**Quantum:** the smallest unit of energy (especially as light) of a given wavelength.

R

**Radiant:** the point in the sky from which a **meteor** shower appears to originate. Each regular shower is named for the constellation from which it appears to radiate; eg. the Leonids in Leo, Perseids in Perseus.
**Radiation:** the direct transmission of heat energy from a hot region to a cooler one; the transmission of energy across space as electromagnetic waves or atomic particles.
**Radioactivity:** the spontaneous disintegration of an atomic **nucleus,** with the emission of an energetic particle or particles.
**Red giant:** a star that is near the end of its life on the **Main Sequence** and has expanded to hundreds of times its original size, at the same time cooling and becoming reddish.
**Red shift:** a **Doppler** shift in the light of a receding object towards the red or longer wavelengths of the **spectrum.**
**Refraction:** the bending of a beam of light when passing from one transparent medium (eg, air) into another (eg. water or glass).
**Regolith:** the powdery 'soil' on the surface of the Moon and some other bodies, caused by continuous meteoritic bombardment.
**Resonance, orbital:** an exact ratio between the revolutions made by one satellite and another; eg. Saturn's moon Enceladus makes two orbits for every one made by Dione— a 2:1 resonance.
**Retrograde motion: rotation** or **revolution** of a body in the opposite direction to that normal in the Solar System; apparent reversal in the motion of a planet in the sky.
**Revolution:** movement of a body around its primary; **sidereal period.**
**Roche's limit:** the distance from the centre of a planet etc. within which a satellite body would be disrupted by **tidal** forces.
**Rotation:** the movement of a body around its **axis.**

S

**Satellite:** a moon; any small body revolving in an **orbit** around a larger body. It is in **free fall,** its speed counterbalancing the pull of **gravity** of the primary body.
**Sidereal period:** the time taken for a planet or other body to make one complete **revolution** of the Sun, or a satellite of its parent planet.
**Sidereal time:** local time measured by the apparent movement of the stars (as opposed to the Sun) on the **celestial sphere,** taking 0 hours as the moment when the **First Point of Aries** crosses the observer's **meridian.** A **sidereal day** is 23h56m4s.
**Solar wind:** the flow of gas and atomic particles from the Sun past the Earth and planets.
**Solstices:** the dates on which the Sun is at its most northerly (22 June) or southerly (22 December) point in the sky, and is at its maximum distance above or below the **celestial equator.**
**Spectrograph:** an instrument for recording a **spectrum** photographically. The recorded image is a spectrogram.
Spectroheliograph: an instrument for photographing the Sun in the light of one particular wavelength.
**Spectroscope:** an instrument used (in conjunction with a telescope) for direct visual observing and analysis of the **spectrum** of a star or other luminous body.
**Spectrum:** the continuous band of colours, displayed in order of **wavelength** from violet to red, into which a ray of white light is separated by a prism or other means.
**Spin-orbit coupling:** an exact resonance between the **rotation** of a body and its **sidereal period;** eg. Mercury rotates three times in two of its 'years'—a 3:2 coupling.
**Star:** a massive, self-luminous gaseous body in the **core** of which energy is produced by **nuclear fusion** reactions.
**Sunspot:** a dark (by contrast), cooler spot on the Sun's bright **photosphere,** caused by a magnetic disturbance.
**Supernova:** a violently exploding star, from which most of the outer layers are blown off, leaving the dense **core.**
**Synchronous orbit:** in the case of an artificial satellite, an orbit in which a satellite remains above one spot on Earth's equator (also known as a geostationary or Clarke orbit, after Arthur C. Clarke who first proposed it). Any two natural bodies can exhibit synchronous rotation if the same face of one is turned permanently towards the other.
**Synodic period:** the time interval between one **opposition** (superior planet) or **conjunction** (inferior planet) and the next.

T

**Temperature:** measure of the heat energy in an atom or molecule of a substance. As the temperature of a **black body** is increased, it radiates light as well as heat; first red, then yellow, white and finally bluish—hence the colours of stars.
**Tektite:** small, glassy objects found only in certain areas of Earth, notably Australia. May originate in the impact of a large **meteorite** or **comet** on Earth, or on the Moon.
**Terminator:** the line between light and dark ('night' and 'day') on any light-reflecting celestial body.
**Terrestrial planet:** basically rocky planets—Mercury, Venus, Earth and Mars.
**Tide:** a bulge raised in the material of a body, or in the oceans of Earth, caused by the **gravitational** influence of a nearby body (eg. the Moon).
**Transit:** the passage of one celestial body across the face of another; the movement of a point on the **celestial sphere** across the observer's **meridian.**

U

**Ultra-violet: radiation,** longer than X-rays but shorter than visible light. It causes sunburn, and is lethal in large doses.
**Universe:** everything that exists; the Cosmos.

V

**Van Allen Belts** (or **Zones**): doughnut shaped belts around the Earth in which electrically charged particles **(ions)** from the Sun are trapped.
**Vernal Equinox:** see **First Point of Aries.**
**Volatiles:** elements that can easily be driven out of a material by heat.

W

**Wavelength:** the distance from the crest of one wave (electromagnetic or other) to that of the next.
**Weight:** the effect on an object of the force of **gravity.**
**White dwarf:** a very small and extremely dense **star.**

X

**X-rays:** very short **radiation,** with a wavelength of about 0.1–100Å.

Y

**Year:** the time taken for Earth to revolve once round the Sun; in normal terms, 365 days (366 in a Leap Year).

Z

**Zenith:** the point on the **celestial sphere** overhead to an observer (90° **altitude**).
**Zodiac:** the twelve constellations lying in a belt about 9° each side of the **ecliptic,** within which all the bright bodies of the Solar System appear to move.
**Zodiacal Light:** a faint cone of light extending to each side of the Sun in the plane of the **ecliptic,** caused by a swarm of dust particles.

# Table of Satellites

| PLANET | SATELLITE | DIAMETER (KM) | MEAN DISTANCE FROM CENTRE OF PRIMARY (KM) | SIDEREAL PERIOD (DAYS) | ORBITAL INCLINATION (DEGREES) |
|---|---|---|---|---|---|
| **Earth** | Moon | 3,476 | 384,000 | 27.32 | 5.15 |
| **Mars** | Phobos | 27 x 21 x 19 | 9,350 | 0.32 | 1.1 |
| | Deimos | 15 x 12 x 11 | 23,500 | 1.26 | 1.6 |
| **Jupiter** | XIV Adastea | 35 ? | 134,000 | 0.30 | 0.0 ? |
| | V Amalthea | 265 x 150 | 181,000 | 0.49 | 0.4 |
| | XV 1979 J2 | 75 ? | 222,000 | 0.67 | 0.0 ? |
| | I Io | 3,640 | 422,000 | 1.77 | 0.0 |
| | II Europa | 3,130 | 671,000 | 3.55 | 0.5 |
| | III Ganymede | 5,270 | 1,070,000 | 7.15 | 0.2 |
| | IV Callisto | 4,840 | 1,880,000 | 16.70 | 0.2 |
| | XIII Leda | 8 ? | 11,110,000 | 240 | 26.7 |
| | VI Himalia | 170 | 11,470,000 | 251 | 27.6 |
| | X Lysithea | 20 | 11,710,000 | 260 | 29.0 |
| | VI Elara | 60 | 11,740,000 | 260 | 24.8 |
| | XII Ananke | 17 | 20,700,000 | 617 | 147.0 (R) |
| | XI Carme | 24 | 22,350,000 | 692 | 164.0 (R) |
| | VIII Pasiphae | 27 | 23,300,000 | 735 | 145.0 (R) |
| | IX Sinope | 21 | 23,700,000 | 758 | 153.0 (R) |
| **Saturn** | S-15 1980 S28 | 200 | 276,400 | 0.60 | 0.0 |
| | S-14 1980 S27 | 250 | 278,800 | 0.61 | 0.0 |
| | S-13 1980 S26 | 300 | 283,400 | 0.63 | 0.0 |
| | S-10 1980 S3 | 200 | 302,800 | 0.69 | 0.0 |
| | S-11 1980 S1 | 70 x 135 | 302,900 | 0.69 | 0.0 |
| | Mimas | 390 | 370,800 | 0.94 | 1.5 |
| | Enceladus | 500 | 476,400 | 1.37 | 0.0 |
| | Tethys | 1,050 | 589,200 | 1.89 | 1.1 |
| | Dione | 1,120 | 754,800 | 2.74 | 0.0 |
| | S-12 1980 S6 | 80 | 754,800 | 2.74 | 0.0 ? |
| | Rhea | 1,530 | 1,053,600 | 4.42 | 0.4 |
| | Titan | 5,118 | 2,444,000 | 15.95 | 0.3 |
| | Hyperion | 360 × 250 | 2,964,000 | 21.28 | 0.4 |
| | Iapetus | 1,440 | 7,124,000 | 79.33 | 14.7 |
| | Phoebe | 150 | 25,920,000 | 550.33 | 150.0 (R) |
| **Uranus** | Miranda | 550 | 130,000 | 1.41 | 0.0 |
| | Ariel | 1,500 | 192,000 | 2.52 | 0.0 |
| | Umbriel | 1,000 | 267,000 | 4.14 | 0.0 |
| | Titania | 1,800 | 438,000 | 8.70 | 0.0 |
| | Oberon | 1,560 | 586,000 | 13.46 | 0.0 |
| **Neptune** | Triton | 5,500 | 354,000 | 5.87 | 159.9 (R) |
| | Nereid | 300 | 5,562,000 | 359.88 | 27.7 |
| **Pluto** | Charon | 1,100–1,500 | 17,000 ? | 6.4 | 57.0 |

((R) = retrograde)

**NB.** Data supplied by different authorities varies; in the above tables, the latest information available (or in some cases an average) has been used. The diameters etc. of the satellites of the outermost planets are not yet known to any degree of accuracy, and are little more than estimates.

# Index

Italic figures indicate that an illustration appears.

Accretion *14 15* 17 26 *28* 67 68
Adams, John 83
Alfven, H 18
Amalthea 74
Amino acids 30
Amor 66
Angular momentum 12 14 18 40
Anorthosite 45
Antoniadi, E.M. 46
Apogee 40
Apollo (asteroid) 66
Apollo missions 44-45 *45*
Asteroids *32* 62 64-67 *65* 85 86
   chondritic 62 85
   exploitation of 86
   Kirkwood Gaps 65
   orbits *10* 11
   Trojan 66
Astrology 35
Atmosphere, Earth 33 *33*
   primordial 18 *28* 29
Atoms 16-17 21 *22* 23
Aurorae, Earth 25
   Jupiter 70 *75*

Bacon, Francis 29
Beer, W 43
Black dwarf 16
Bode, J.E. 64 80
Brady, J.L. 85
Brahe, Tycho 35
British Interplanetary Society 85 86

Callisto 70-72 *72*
Caloris basin 46 *49*
Carbonaceous chondrite 15 16 62 67
Cassegrain, N. 39
Cassini, G.D. 42 50 56 76
   Division 76 *76 77 78* 79
Cells, first 30
Ceres *65* 67
Chamberlin, T.C. 12
Charon 84 *85*
Chiron 85
Christy, James 84
Chromosphere 20 *21* 24
Chryse Planitia *57 59* 60 *60*
Colonies, space 86
Comet 11 *66* 67
   Halley's 64 67

Convection 17 *21* 23 26 29
Copernican system *37*
Copernicus, Nikolaus 35
Corona 20 *21* 24 25
Cosmogony 14
Cratering, early 18 *28* 29 72
Craters, impact *44-45* 67 70-72 *72* 79 *79*
   lunar 44-45 *44-45*
   volcanic 44 *44*
Curie point 29

Darwin, Charles 34 *35*
Darwin, George 40
Davis, Raymond 23
Deferents 35 *36*
Deimos *57 59* 62 67
Descartes, René 14
Deuterium *22*
Differentiation, chemical 17 *28*
Dinosaur 35 *31*
Dione 76 79 *79*
DNA 30
Dollfuss, A. 79
Dollond, John 39

Earth, asthenosphere *28*
   atmosphere 33 *33*
   climate 33
   continents 26 *28* 29
   core 26 *28*
   crust 26 *28*
   eras 30-34 *30-31*
   erosion 33
   evolution 26-33 *28*
   from space *32*
   lithosphere *28* 29
   life, origin of 30-31 *30-31*
   lithosphere *28* 29
   magnetic field *25* 26 29 30
   magnetosphere 25 *25* 70
   mantle 26 *28* 29
   mesosphere *28*
   oceans 33
   orbit *10*
   physical data *26*
   ring 82
   sky 8 9 *9*
   tides 33
   weather 33 *33*
Earthlight *41*
Eclipse, lunar 40 *41*
   solar 20 *21* 40
Ecliptic *38* 40
Einstein, Albert 85

Ejecta blanket, Mars *60*
   Mercury 46
   Moon 45
Enceladus 76 79
Epicycles 35 *36*
Equator, celestial *38*
Eros *65* 66
Europa 72 *73*
Evolution, Theory of 34 35

First point of Aries *38*
Flamsteed, Rev. John 39
Fraunhofer, J. von 20
   lines 20 *22*
Fusion, nuclear 18 20 *22* 23
Future exploration 86-87 *87*

Galilei, Galileo 35 *38* 42 56 76 85
Galileo probe 7 74
Gamma rays *22* 23
Ganymede 72 *72*
Gauss, Karl 64
Gilbert, Dr. William 26
Grains, dust *13* 14 *14 15* 16 17
Gravity 38 40
Greenhouse effect 52 54 61 86

Hadley Cells 33 *33*
Half-life 17
Halley, Edmond 39 67
Helium 20 *22*
Helmholtz, Hermann von 16
   contraction 16 *16* 20
Hencke, K.J. 64
Hermes 66
Herschel, John 20
   Sir William 20 39 76 80 *80*
Hevelius, Johann 42
Hidalgo *65* 66
High luminosity phase 17
Hipparchus 35
Homo erectus 34
   sapiens 34
Hooke, Robert 68
Hoyle, Sir F. 12
Hunt, Dr. Garry *71* 79
Huxley, T.H. *35*
Huygens, Christian 38 56 76 *76*
Hydrogen 20 *22*

Iapetus 76 79
Icarus 66
Infra-red 16 *22* 23
International Astronomical Union 43 60

Io 72-74 *73 74 75*
   volcanoes 72-75 *73 75*
Ionosphere Earth 25 33
   Venus 54
Ions 18 *22* 23 24
Isotopes 15 17

Jeans, Sir James 12 20
Juno *65*
Jupiter *65* 68-75
   atmosphere 68 *69*
   belts and zones 68 *69* 70 *70* 71
   composition 68 *69*
   core 68 *69*
   density 68
   Great Red Spot 68 *69 70* 71
   life 70
   magnetic field 70
   orbit *10*
   physical data *68* 68-70
   poles *71*
   radio waves 70
   rings *71* 74 *75*
   satellites 70-75 *72-75*
   spots 70 *71*
   temperature 68

Kant, Immanuel 12
Kepler, Johannes 35 *38* 64
   Laws 38 82 84 91
Kowal, C.T. 85
KREEP norite 45
Kuiper, G.P. 12 14 76 80 84

L-5 Society 86
Lagrangian points 66 82
Laplace, Pierre de 12
Lawton, A.T. 85
Le Verrier, Urbain 83 85
Lowell, Percival 56 84
Lunan, Duncan 62
Lyell, Charles 34 *34*
Lyot, B. *21 47*
Lyttleton, R.A. 84

Mädler, J. von 43
Magnetosphere 25
   Earth 25 *25* 26
   Jupiter 70
   Mercury 46 *47*
   Saturn 79
Mariner probes
   2    52
   4    59
   5    52

   6    59
   7    59
   9    59 60
   10   46 *48 49* 52 *53*
Mars 56-63 86 *87 88-89*
   atmosphere 56 *58* 59 60 61
   canals 56 59
   clouds 59
   core *63*
   craters *58* 59 *59* 60 *60*
   life 56 59 62 63
   map *59*
   oppositions 56 *57*
   orbit *9 10* 56 *57*
   physical data 56
   polar caps 56 *58* 59 60 61
   ring 82
   rotation 56
   satellites *57* 59 62 *62-63* 67 82 *88-89*
   temperature 59
   terraforming 86 *88-89*
   terrain *58* 59 60 *60* 61 *62-63*
   volcanoes *58* 59 60 61 *62-63*
Mass-driver 86 *88-89*
McLaughlin, D.B. 61
Mercury 46-49 *47 48 49* 86
   appearance from Earth *47*
   core *47*
   craters 46 *48 49*
   magnetic field 46 *47*
   magnetosphere 46 *47*
   orbit *10 47* 85
   physical data *46*
   rotation 46 *47*
   temperature 46
   terraforming 86
   terrain 46 *48 49*
   transits 50
Meteor 11 *66* 67
Meteor shower *66* 67
Meteorite 11 15 67
   Allende 15
Meteoroid 11
Micrometeorite 67
Milky Way 9 *11* 14
Miller, Stanley L. 30 70
Mimas 76 79 *79*
Molecules, organic 14 30 70
Moon, the 8 *38* 40-45 *41-45* 86
   age 40
   bases 86
   core 45
   craters *41 42 43* 44 *44-45*
   eclipses 40 *41*
   formation 40

   libration 44
   map, far side *43*
      near side *42*
   maria 44
   orbit 40 *41*
   phases 8 *9* 40 *41* 42
   physical data *40*
   probes 44 *44-45*
   terraforming 86
Moulton, F.R. 12

Nebula, cocoon 16 18
   primordial 12-18 *13 14 16-17*
Nebular hypothesis 12
Neptune 82-83 *82-83*
   orbit *10* 84
   physical data *80*
   satellites 83 *82-83*
Nereid 83 84
Neutrino *22* 23
Newton, Sir Isaac 20 38 39
Nicholas of Cusa 9
Nodes *38* 40
   precession of 40

Observations, early 8
O'Keefe Dr. J.A. 82
Olbers, Dr. W.M. 64
Olympus Mons *58 59* 60 61
Oort, Jan 67
Oparin, A.I. 30
Orbiter satellite *41 43* 44 *45*
Orrery *11*
Outgassing *28* 29 61

Pallas *65*
Pangaea 29 *29* 31
Parkinson, Dr. R.C. 86 *87*
Perigee 40
Phobos *57* 59 62 *62-63* 67 82 *88-89*
Phoebe 79
Photons 20 *22* 23
Piazzi, Professor Guiseppe 64
Pickering, Professor W.H. 84
Pioneer probes 70 79
   Venus 52 *53* 54
Planet X 84 85
Planetesimals 12 *15 16-17* 17 18 26 *28*
Plasma 24 25
   tail *66* 67
Plate tectonics, Earth *28* 29 *29*
   Mars 61
   Venus *53*
Pluto 84-85 *85*
   diameter 84

orbit *10* 84
physical data *84*
satellite 84 *85*
Pole, celestial *38*
Positron *22*
Pouillet, C.S.M. 20
Power stations, solar 86
Precession *38* 40 85
Proton proton cycle 20 *22*
Protoplanets *15* 17 *19* 26
Protosun 12-19 *14 16-17 19* 26
Ptolemaic system 35 *36*
Ptolemy 35 *36*

Radiation 21 23 67
Radiation pressure 14 18
Radioactive decay 17 26 *28* 29
Radiolaria 26 82
Radio waves 23 *23*
Regolith 45
Relativity, Theory of 85 87
Rhea 76 79 *79*
Riccioli, G.B. 42
Rilles 45
Roche limit 76 79
Rocks, Earth: igneous 31
     metamorphic 31 33
     sedimentary 31 33
  Mars *58* 61
  Moon 15 40 45
  Venus 52 *52*
Rotation, synchronous 40 84

Sagan, Dr. Carl 52 61 70 86
Saros cycle 40
Saturn 76-79 *76 77 78 front end-paper*
  atmosphere 76
  belts and zones *76-77*
  core *77*
  density 76
  magnetosphere 79
  orbit *10*
  physical data 76 *76*
  rings 76-79 *76 77 78*
  satellites 76-79 *76 79*
  spots *78* 79
Schiaparelli, Giovanni 46 56
Schröter, J.H. 43
Shooting stars *66* 67
Skylab 25
Solar cycle 24 *25*
Solar System *10 11*
Solar wind 25 *25 28* 67
Spacelab 86
Space Shuttle 39 86

Space Telescope 39 *39*
Spectroheliograph 20
Spectroscope 20 *22*
Spectrum 20 *22-23* 38
Stars, infra-red 16
  T-Tauri 18
  variable 18
Stratigraphical column *30-31*
Sun *19* 20-25 *21-25*
  as red giant 87 *back end-paper*
  chromosphere 20 *21* 24
  core 20 *21* 23
  corona 20 *21* 24 25
  coronal holes 25
  faculae 20 *21* 24
  fibrils 24
  filaments 24
  flares 24
  granulation 23 24
  magnetic field 23-25 *24 25*
  photosphere *21 22* 23 24 *25*
  physical data 20 87
  plage 24
  poles *21* 25
  prominences 20 *21* 24 *24*
  radio waves 25
  rotation 24
  spicules 24
  supergranulation 24
Sunspots 20 *21* 23 24 *24*
Supernova 12 *13* 15 16

Tektites 82
Telescope, Cassegrain 39
  Mt. Palomar 39
  Mt. Wilson 39
  reflecting *38* 39
  refracting 35 *38* 39
  Space 39 *39*
Terraforming 86 *88-89*
Tethys 76 79 *79*
Tides 33 39 40
Titan 38 79 *79* 87
Titius-Bode Law 64 83
Tombaugh, Clyde 84 85
Triton *82-83* 83 84
T-Tauri wind 18 *19* 81

Ultra-violet *22* 23
Umbriel *81*
Universe, physical data 12 *12*
Uranus 80-82 *81*
  core 80
  physical data *80*
  rings 80 *81* 82

satellites 80 *81*
seasons *81*

Valles Marineris *58 59* 60 61 87 *88-89*
Van Allen. Dr, J.A. *25*
  belts 25
Venera probes 52 *52* 54
Venus *47* 50-55 *51-55* 86
  airglow 50
  albedo 50
  appearance from Earth 50 *51*
  Ashen Light 50 54
  atmosphere 50 *51* 52 54
  clouds 50 *51* 52 *53* 54
  core 52
  craters 52 *53*
  ionosphere 54
  magnetic field 50 54
  map *53*
  orbit *10 47* 50
  physical data *50*
  rotation 50 52
  temperature 50 *51* 52 54
  terraforming 86
  terrain 52 *53*
  transits 50 *51*
  volcanoes 52 *53 54-55*
Vesta *65*
Viking probes *57 58* 60 61 63
Volcanoes, Earth *27 28* 29
    Io 72-75 *73 74 75*
    Mars *58 59* 60 61 *63*
    Moon 43 44 *44*
    Venus 52 *53 54 55*
Voyager probes 70-74 76-79 80
  *front end-paper*
Vulcan 85

Wegener, Alfred L 29
Weizacker, C. von 12
Wells, H.G. 59
Wildt, Rupert *69* 80
Wilson, Alexander 20
Wilson, L.A. 80
Wolf, Professor Max. 66
Wollaston, W.H. 20

X-rays *22* 23 25

Zodiac 8 *38*
Zodiacal Light *19* 25

# Acknowledgements

All photographic illustration supplied by NASA/JPL except
where stated:
Radio Times Hulton Picture Library, Orrery page 11
National Portrait Gallery, pages 34-5, 80
Novosti Press Agency, page 52

The author gratefully acknowledges help from the following:
*Astronomy*
British Broadcasting Corporation
British Interplanetary Society (and *Spaceflight*)
Damian Grint
Dale Kornfeld (NASA)
*Future Life*
Dr. Garry Hunt
*Icarus*
Jet Propulsion Laboratory (JPL)
*Journal of the British Astronomical Association*
Bryan Lee
Chris Morgan
National Aeronautics and Space Administration
*Nature*
*New Scientist*
*Planetary Report*
Ian Ridpath
*Scientific American*
*Technology Review*

The following societies and organisations are concerned with
astronomy, planetary exploration, astronautics and related subjects.
All are international, and open for membership.

British Astronomical Association: Secretary, Burlington House,
    Piccadilly, London W1V 0NL
British Interplanetary Society: 27/29 South Lambeth Road,
    London SW8 1SZ
Junior Astronomical Society: Secretary V L Tibbott, 58 Vaughan
    Gardens, Ilford, Essex LG1 3PD
Association in Scotland to Research into Astronautics (ASTRA):
    49 Almada Street, Hamilton, Lanarkshire NL3 0HL
The Planetary Society: P O Box 3599, Pasadena, California 91103,
    USA
National Space Institute: P O Box 1420, Arlington, Virginia 22210,
    USA
L-5 Society: 1060 E. Elm, Tucson, Arizona 85719, USA